Bataille's Peak

Bataille's Peak

Energy, Religion, and Postsustainability

Allan Stoekl

UNIVERSITY OF MINNESOTA PRESS
MINNEAPOLIS – LONDON

The University of Minnesota Press gratefully acknowledges the financial assistance provided by the Penn State French Department for the publication of this book.

Part of chapter 2 was previously published as "Excess and Depletion: Bataille's Surprisingly Ethical Model of Expenditure," in *Reading Bataille Now,* ed. Shannon Winnubst (Bloomington: Indiana University Press, 2007), 252–82.

Published by the University of Minnesota Press
111 Third Avenue South, Suite 290
Minneapolis, MN 55401-2520
http://www.upress.umn.edu

Library of Congress Cataloging-in-Publication Data

Stoekl, Allan.
Bataille's peak : energy, religion, and postsustainability / Allan Stoekl.
p. cm.
Includes bibliographical references and index.
ISBN: 978-0-8166-4818-4 (hc : alk. paper)
ISBN-10: 0-8166-4818-2 (hc : alk. paper)
ISBN: 978-0-8166-4819-1 (pbk. : alk. paper)
ISBN-10: 0-8166-4819-0 (pbk. : alk. paper)
1. Bataille, Georges, 1897–1962. I. Title.
B2430.B33954S76 2007
194—dc22 2007014874

Printed in the United States of America on acid-free paper

The University of Minnesota is an equal-opportunity educator and employer.

12 11 10 09 08 07 10 9 8 7 6 5 4 3 2 1

Contents

Acknowledgments

I especially thank those who read the manuscript of this book, or parts of it, and gave me much useful feedback: Jeff Pruchnik, Shannon Winnubst, and Neil Hertz. Alphonso Lingis, as always, provided inspiring intellectual—and moral—support. Many thanks as well to the two anonymous reviewers who evaluated this book for the University of Minnesota Press and to my wonderful editor, Richard Morrison.

I also greatly appreciate the generosity of those at the Liberal Arts College here at Penn State who provided the necessary sabbatical.

And—speaking of generosity—how do I acknowledge that of Nan Moschella?

Just as I started writing, my mother, Mary Ann Steinfort Stoekl, passed away. This one's for you, Ma. I know you'll find a place to stash it.

Introduction

On Shortage, Excess, and Expenditure

At the end of the twentieth century, we were regaled with arguments concerning history: it had ended, we were told. The Franco-Russian philosopher Alexandre Kojève had been right when he argued in the 1930s and 1940s that at a certain point *nothing new could happen.* Human liberation, inseparable from human labor and the progress of philosophy, had ended; a state in which freedom was attained through the recognition of the freedom of the other was definitive. From now on a State that implemented that freedom was all that could be postulated; all else would constitute a fall backward into a historical movement that had, for all intents and purposes, ended. Of course things would still "happen"—tsunamis, earthquakes, famines—but the essential narrative was over. "Man's" labor was complete—labor in the larger sense, in the sense of the construction of the meaning of the human and the concomitant end of the human (as a process of development). The accession to the end might be piecemeal—not all societies would arrive at it simultaneously—but once arrived at, it would be definitive.[1]

Barely a few years into the new century, many have concluded that that "posthistorical" ideal is radically insufficient. Kojève's model was structured in such a way that major historical events and changes would always seem minor—Kojève himself dismissed the importance of World War II, arguing that it was little more than a preparation for the final stage of a grand synthesis of American capitalism and Soviet collectivism (the ultimate state-sponsored consumer society, in other words). But something else—a new event, if we can call it that—has appeared that puts into question the very possibility of ending history and above all ending it because the "labor of the negative" has been completed. The very centrality of human labor is one of the things most in question.

In short, energy has been rediscovered. In the 1970s and very early 1980s, first world society was made acutely aware of energy, its limited supplies, and the consequences of energy shortages. A U.S. president (Jimmy Carter) even based his central policies on the idea that energy sources (fossil fuels) were scarce and could only grow scarcer in the coming years. He wore a cardigan (this is all that seems to be remembered about him) and, from the Oval Office, warned Americans that they would have to tighten their belts, turn down the thermostat, find alternative energy sources, get ready for dire, gray days, and gird for the "moral equivalent of war." He was, of course, brusquely turned out of office and replaced by a president (Ronald Reagan) who cheerfully answered that the "free market" would take care of energy supplies forever. Luckily for him, the quantities of fossil fuels available shot up in the mid to late 1980s and throughout the 1990s because of conservation measures set up under Carter and because a few new sources of oil (from the North Slope of Alaska and from underwater fields in the North Sea) became available. By the late 1990s, oil was down to $10 a barrel.

As I write this, in 2006, even mainstream news sources have become aware that fuel supplies are fundamentally limited. Competition from China, India, and other developing nations, and the decline of American wells, has led to a situation in which there is virtually no excess oil production capacity anywhere in the world. As Gerald F. Seib puts it in the *Wall Street Journal:*

> From Iraq to China, from the Gaza Strip to Iran, the biggest foreign-policy problems of the summer [of 2005] all are setting off the same alarm: It is imperative for the US to become more energy independent.
>
> But that, of course, is precisely what Washington's policymakers have been unable, or unwilling, to accomplish. Instead, America's exposure to trouble in the world's volatile oil-producing regions actually is on the rise, even as the summer driving season heads toward its climax with oil near a once-unthinkable $65 a barrel. In brief, while the 20th century was the century of oil, the 21st already is unfolding as the century of whatever follows oil, or the century of fighting over what's left of oil—or both.[2]

The labor of the construction of civilization is not over, in other words, history is not at an end, because labor itself is not autonomous: you can't work or produce anything if you don't have the fuels (the sources of energy) to do it. The great myth that Man "forms himself" by forming, and transforming, brute matter is over. The idea that Nature is dead is over because fossil fuels were not made by Man, they were only extracted by "him." They are brutally natural, and their shortage too is a natural shortage (their *lack* is natural). And when a profound, irremediable shortage of those fuels supervenes, history opens back up. History will not, as some critics of the "end of history" thesis claimed, return merely as localized struggles and revolts that put the superpowers on the spot. Instead, History now is the fight for a resource that will allow History as we have come to think of it—the flourishing of civilization and the establishment of the definitive dignity of Man—to continue and triumph. No one yet wants to think about how History should continue in the absence of an adequate supply of fossil fuels. It is too horrible to think about. Human die-off is quite natural, but it also constitutes an incontrovertible historical event.[3] With the finitude of cheap energy, alas, the end of history is itself finite. But how do we think the end of the end of history?

Now along with a permanent energy crisis, or rather a permanent shortage of cheap fuel supplies, we face another crisis: a permanent religion crisis. It seems as if energy and religion are inseparable issues. On the simplest level (and this was already apparent in the 1990s), the decline of the secular dream of the end of history explained rationally and scientifically—e.g., the decline of Marxism *and* capitalism—resulted in a turn, in many parts of the world, to other models of strategy and solidarity. If the ultimate secular, rational models of community building and future understanding failed, they could only be replaced with more traditional and less rational models: Christianity, Islam, and other religious creeds. At first it seemed that these particular modes of belief stood little chance of challenging the mainstream new world order, the order of the termination of history in secular citizenship and the universal ideal of proletarian solidarity or contented, suburban moneymaking.

The first crack in the ideal of posthistorical reason was to be found in the decline of Marxism. Marxism too posited an end of history, one that was definitive, secular, rational, grounded in the physical and moral contentment of Man. But Marxism was the canary in the coal mine, so to speak, because its decline was due to an energy crisis, the first to shock the world since the crises of the late 1970s. Marxism collapsed because its great, worldwide patron, the Soviet Union, collapsed, and the Soviet Union collapsed because it could no longer support itself by selling its oil profitably on the world markets.[4] It was driven into the ground by Saudi Arabia, which in the late 1980s produced so much oil that the world markets became flooded. And at that very moment, the Soviets discovered that their oil production had peaked and was entering into decline. The Soviets quickly went out of business, as did their ideology. (The Saudis knew what they were doing.)

The great irony is that religion came to the fore in the very countries whose vast production of fossil fuels had made the Soviet system untenable. The Islamic countries of the Middle East were the producers of the fuels that the West needed to continue its individualist lifestyle. But they got little for what they sold—just diminishing fuel reserves, falling oil prices (since they were forced to compete with each other), and increasingly impoverished populations. The humanism of the posthistorical era, be it American or Soviet, was over, replaced by a theological realm that recognized no difference between religious strictures and the laws of human comportment. In this view, neither human labor nor fossil fuel ultimately made possible the world's survival; instead, God was the ultimate referent, man and energy alike sinking into insignificance before him. Energy for religion was nothing more than a resource to be sold to enable not so much a higher standard of living (the posthistorical imperative) but a more perfect level of worship.

Thus the standoff of the early years of the twenty-first century. Secular humanist countries, practicing a rigorous separation of church and state, crave oil because their lifestyle depends on it: they engage in a lavish expenditure inseparable from the wanton waste of the easily refined energy available, in concentrated form, in fossil fuels. Many of the regions that provide these fuels have turned to a religion that is, in principle at least, indifferent to the fossil fuel lifestyle and to the cult of the human. Many in the high-consumption world have turned to fundamentalist religion as well, perhaps in reaction to the embrace of fundamentalism in the fuel-exporting world.

Fuel production, fuel consumption, conflict over fuel; energy shortage, religion surplus in reaction to it; resource wars, religious wars, history after the supposed end of history. As fuel reveals its finitude, we come to recognize our dependence on it and our dependence on others who affirm a religious culture that survived and flourished in the profound absence of fossil fuel.[5]

There is, however, a deeper connection between energy and religion. Energy is not just a commodity to be measured, stockpiled, sold, consumed, wasted. And religion is not just a method of resisting a relentless movement of production-consumption, nor is it merely a means of providing a stable alternative (God) that can ground society in the absence of (or against) the delusive subjectivity of the "age of the world picture." Energy may in fact be a profoundly religious issue—energy in its vastness, its violence, its defiance, its elusiveness, its expenditure. And religion may be an event not of the establishment of God, or of his patronage of humankind, but of his death, his void at the peak of values and purposes. God's death, in effect, may very well be inseparable from the movement of the violent expenditure of energy, *all types* of energy.

The French writer Georges Bataille (1897–1962) put forward a social model that sees religion and human existence as inextricable, and the religious experience—sacrifice—as entailing the profligate wastage of energy. But therein lie the central questions: Which religion? And which energy?

This book is about Bataille's take on these issues and my version of what Bataille's take would be if it were extrapolated to the twenty-first century. Bataille died a long time ago, ages ago it seems, but one can perhaps rewrite him, all the while recognizing certain limitations of his approach, in an attempt to understand the *possibilities* of the future in a post–fossil fuel era. That's what I try to do in this book. Bataille is hardly the last word on anything, but he is rare—in fact, unique—among twentieth-century thinkers in that he put energy at the forefront of his thinking of society: we are energy, our very being consists of the expenditure of quantities of energy. In this Bataille anticipates scientists like Howard Odum, who in a very precise way calculate the amounts of energy that go into a given product, a given lifestyle, and so on (and calculate as well how we can work to make the processes of production and consumption more efficient, given the scarcity of recoverable fuels). But Bataille is about more than simply quantifying energy; indeed, his approach both sees energy at the basis of all human activity, of the human, and puts into question the dominion of quantifiable, usable energy. That is precisely where religion comes in, since God, or religious "experience," entails not purposive activity—the kind that would involve energy supplies quantified and then used with a goal in mind—but rather activity of the instant that leads nowhere, has no use, and is unconditioned by the demands of anyone or anything else: sovereign, in Bataille's sense. Such sovereign activity involves an energy resistant to easy use—the unleashing of an energy that is characterized (if that is the word) by its insubordination to human purposes, its defiance of the very human tendency to refine its easy use.

My consideration of Bataille, then, will necessarily involve a critique of the notions of energy and religion that characterize our epoch—an epoch in for some interesting times as cheaply available energy from fossil fuels grows scarcer and scarcer. It will attempt to imagine how other notions of energy and religion will provide an alternative means of living in an era in which the *truth* of fossil fuel, and revealed religion, comes into question. Another model of spending, based on what Bataille called an "economy on the scale of the universe," seems appropriate at a time when a certain human profligacy has revealed itself to be an ecological and cultural dead end. Bataille's importance, however, stems from the fact that he puts forward a model of society that does not renounce profligate spending, but affirms it. What is affirmed, however, is a different spending—a different energy, a different religion—and that difference perhaps means the difference between the simple meltdown of a civilization and its possible continuation, but on a very different "scale."

On the other hand, an ever more counterproductive orientation will assert itself in the years ahead. Such an orientation sees energy as an adjunct of, at best, a certain humanism: we spend to establish and maintain our independent, purpose-driven selves, our freedom as consumers, spenders of certain (rather lavish, given available reserves) quantities of refined energy, This model is doubly humanistic in that not only is the beneficiary the "free" self of Man; the human spirit itself is incessantly invoked to get us out of the jam. We are told over and over again that the human mind alone produces energy: when reserves are short, there is always a genius who comes along and devises some technology that turns things around, makes even more energy available, and so on.[6] Technology transcends energy, in other words, and reflects the human mind's infinite ability to derive energy from virtually nothing. We always find more efficient ways to derive energy from available fuels, and in doing so, we always are able to produce more fuel to produce more and higher quality energy. James Watt's steam engine was first used to drain coal mines, producing more coal, which in turn could be used by more (and more efficient) steam engines to produce transportation (steam trains), electricity, and so on. And petroleum, an even more productive and efficient source of energy, replaced coal, and it will no doubt soon be replaced by something else, yet to be discovered. At this point we move from a historical account to a kind of uncritical faith in the capacity of human genius.

Fossil fuels, then, entail a double humanism: they are burned to serve, to magnify, to glorify the human or (what amounts to the same thing) the human in the automobile ("freedom," "happiness," etc.) as transcendental referent, and they are produced solely through the free exercise of the mind and will.

One can argue that the religion that confronts the fossil fuel–driven civilization of Man is equally grounded in the demands of a human subjectivity. People demand salvation, an ultimate purpose for which they are consuming so much fuel: I spend, or waste, so that I will ultimately be saved. Conversely, energy inputs are available because God has blessed me with them; the faithful are rewarded with a healthy, fertile, and energy-rich environment. God is the ultimate meaning of all that I think and do. There is no distinction between my personal belief and belief sanctioned by society, derived from a literal reading of a holy Book. In order to give this version of religious belief even more authority, law is grounded not in man but in God himself; literalism serves as a satisfying alternative to humanism.

Against this energetico-theological model is arrayed an ecoreligion, one that would defy the "comfortable" or "free" (and nonnegotiable) lifestyle of consumerist humanism, not through a recognition of the literal truth of the divine Word but through a religiously inspired cult of austerity, simplicity, and personal virtue. Such a cult refuses certain basic human urges to consume or destroy, and in the process involves the affirmation of yet another humanism (the self as virtuous in its austerity) and, after consumer profligacy, yet another model of nature as a standing reserve to be protected largely for its value to Man.[7]

Fossil fuel civilization, then, and its antitheses, or antidotes. Man and/or God as ultimate referent: a couple we can expect to hear more from in the coming years. Bataille poses a very different model of the interrelation of energy and religion. This is not to say, however, that the spending Bataille examines somehow replaces or is more fundamental, more originary, than the consumerist or religious models it confronts. Instead we might say that Bataille's vision is the underside, the ungraspable double that has been there from the first effort of the human, that it asserts itself precisely at the moment in which the finitude of the human manifests itself through the recognition of the limits of fossil fuel energy itself. Bataille's energy and religion are not an alternative; they promise nothing for the future, certainly no salvation, although their aftereffect may entail a future more livable—by whom?—than that promised under the signs of God or Man.

Bataille's energy is inseparable from that which powers cars and raises elevators, but it is different as well. It is excess energy, and in that sense it is left over when a job is done, when the limits of growth are reached, or, in the current situation, when fossil fuels themselves reveal their profound limitations. Bataille's energy is a transgression of the limit; it is what is left over in excess of what can be used within a fundamentally limited human field. As such, it is quite different from what *can* be used: it is not just left over in the sense of not being consumed; it is fundamentally unusable. At the point at which quantification reveals its finitude, energy asserts itself as the movement that cannot be stockpiled or quantified. It is the energy that by definition does not do work, that is insubordinate, that plays *now* rather than contributing to some effort that may mean something at some later date and that is devoted to some transcendent goal or principle. It is, as Bataille reminds us a number of times, the energy of the universe, the energy of stars and "celestial bodies" that do no work, whose fire contributes to nothing. On earth, it is the energy that traverses our bodies, that moves them in useless and time-consuming ways, that leads to nothing beyond death or pointless erotic expenditure, that defies quantification in measure: elapsed moments, dollars per hour, indulgences saved up for quicker entry into heaven. Energy is expended in social ritual that is pointless, that is tied not to the adhesion of a group or the security of the individual but to the loss of group and individual identity—sacrifice.

Bataille's religion is thus inseparable from Bataille's energy. Sacrifice is the movement of the opening out, the "communication," of self and community with death: the void of the universe, the dead God. These are not entities that can be known or studied, but sovereign moments, moments of unconditional expenditure. This entails the expenditure of certainties, of any attempt to establish a transcendent, unconditioned meaning that grounds all human activity, a referent such as Man or God. Precisely because it really is unconditioned, this meaning—God, if you will—is sovereign, dependent on nothing, and certainly not on Man and his petty desire or demands. Religion, in the orgiastic movement of the body, is the loss of transcendent meaning, the death of God as virulent force, the traversing of the body by an energy that overflows the limits it recognizes but does not affirm. If there is community, it is the unplanned aftereffect and not the essential meaning of this energy, of this movement of the death or void of God.

Thus ethics for Bataille, the community, and its meaning and survival are aftereffects of the expenditure of the sacred. Bataille's theory is profoundly ethical but only in the sense that the instant of preservation, of meaning, of conservation, of knowledge, is the unforeseen offshoot of another movement, that of the drive to spend without counting, without attempting to anticipate return. To deny the ethical moment, the moment in which conservation and meaning are established only the better to affirm the destruction of expenditure, is to relegate that destruction to the simple, homogeneous movement of the animal, unaware of limit, meaning, and purposive act. Expenditure, in other words, is not the denial of the human, its repression, but instead its affirmation to the point at which it falls: the sacrificial act, the recognition of an energy that does not do "work" for the maintenance of the human, is the affirmation of a God who is not the slave of the human. It is the impossible moment in which awareness doubles the unknowable loss of energy and the virulence of a God who disbelieves in himself.

The ethics of Bataille, then, entail a vision of the future in which the "left-hand sacred," the sacred of impurity, of eroticism, of the radically unconditioned God, spins off a community in and through which expenditure can be furthered (a community of those with nothing in common). Not nuclear war, but the channeling of excess in ways that ensure survival so that more excess can be thrown off. And (one can continue along these lines) not generalized ecocide, but an affirmation of another energy, another religion, another waste, entailing not so much a steady state sustainability (with what stable referent? Man?) but instead a postsustainable state in which we labor in order to expend, not conserve. Hence the energy, and wealth, of the body—the energy of libidinous and divine recycling, not the stockpiled, exploited, and dissipated energy of easily measured and used fossil fuels.

This book has two goals: in the first part, to sketch out Bataille's positions on energy expenditure, religion of and against the Book, and the city; in the second, to extrapolate from those positions and consider current questions of energy use and depletion, religious literalism and fervor, and urban "life." Urban space is a crucial problem for Bataille in that for him the city is the privileged locus of the physical and geographical elaboration of the sacred: either the right-hand sacred of concentration, hierarchy, and God as repressive force or the left-hand sacred of dispersal, the fall of meaning and sense, and God as figure of the sovereign expenditure of authority.[8] The city is, finally, the locus of concentration in and as the modern, and any consideration of a transition from an energy-religion complex of Man to one of the death of Man entails a reconsideration of the city as spatial and economic structure: a reconsideration that proposes not just energy efficiency and sustainability, but those elements as aftereffects of a more profound burn-off.

The first four chapters of the book are therefore concerned first with the intellectual antecedents of Bataille—Giordano Bruno and the Marquis de Sade—in whose writings a profound connection between untamed matter-energy and religion as the dispersal or death of God is already evident (and who are rewritten in significant ways by Bataille); then, successively, with Bataille's theories of energy, religion, and the city. In these chapters I am concerned with the ways in which Bataille specifically established his theory against certain positions that had undeniable force: in the case of energy, against a version of scarcity that implied only conservation and utility (the Protestant ethic, communism); in the case of religion, against immutable doctrines of the Book (the Bible, Kojève, Sade); in the case of the city, against the image of the city as method of human concentration and hierarchy through religious-secular monuments and mechanized transport.

The second part of the book, my rewriting of Bataille in light of the current energy-religion articulation, follows the same order: energy (chapter 5), religion (chapter 6), the city (chapter 7). Here, however, I want to consider how Bataille's theory helps us understand energy and reconceive it (this entails a reading of Heidegger on technology as well); how his critique of religion lends itself to a reconsideration not so much of traditional religion, which Bataille himself has already carried out, but of contemporary humanist and/or fundamentalist ecotheology (doctrines that bear an eerie resemblance to the secular cults of the Book demonstrated by Sade and Kojève); and finally how Bataille enables us to think about the death of God in the post–fossil fuel city, the city that will likely be left over after the dreams of inanimate resource-rich wastage in perpetuity have revealed, quite palpably, their limitations. In each case, I attempt to elaborate a Bataillean ethics based on certain simple premises: that fossil fuel energy is entering the early stages of depletion and that it is time to think the consequences of another type of energy and another kind of relation to recycling and to matter in general; that humanist ecoreligion or theocratic ecofundamentalism is not the inevitable result of the decline of a fossil fuel economy and the return of a solar or (more generally) renewable energy economy; that, in other words, a critique of modern, humanist, or antihumanist religion leads not to the triumph of Man but to his death in glory or agony (hence to another religion, one on the "scale of the universe"); and finally that the city is the privileged space of this social transformation but in the future the city must be conceived as a topography of spectacular energy expenditure in the largest, Bataillean sense of the term, rather than as a mere locus of energy use and conservation.

I am above all concerned with strategies that will allow us to elaborate Bataille's ethics. As I have noted, these ethics entail a certain blindness: the left hand does not always know what the right is doing, or in a Nietzschean formulation, one loves a profound ignorance concerning the future. The future, I argue, is fundamentally resistant to planning; blind expenditure entails not an obsessive and centralized prognostication, authored by a head that is always the supreme metonym and referent of social intelligence, but rather the playing out of aftereffects in which social practices may very well "save the earth" in spite of themselves (save it not for conservation but for lavish consumption). An economy on the scale of the universe implies an earth on the scale of the universe. Recycling, for example (as I try to show in chapter 5, in my presentation of Agnès Varda's film *The Gleaners and I*), is not merely a question of a new, slightly more benign form of maintaining a standing reserve; on the contrary, it is the orgiastic movement of the parody of meaning, of the expenditure of the energy of meanings and of physical and social bodies, an ethics (and aesthetics) of filth, of orgiastic recycling. Similarly, a religion on the scale of the universe means one that rejects the inevitable: if renewable energy in the past has always spawned some version of feudalism and fundamentalism, our profound ignorance of the future precludes this simple and self-defeating certainty.[9] A critique of humanism or fundamentalism means the refusal to see God or Man as the ultimate signified before which all (energy) slaves bow. The expenditure of Kojève's Hegel, after all, means the loss of all certainty, all dialectical and labor-oriented modes of the establishment of (terminal) meaning, history, and value. The future for that reason is not necessarily a reverse replay of military-labor-philosophical history, in which a mass die-off is accompanied by gradual cultural collapse. A future, renewable energy society—one based on the glorious expenditure of unrefinable energy and not its obsessive and impossible conservation—means a muscle-based, human-powered, but literally postmodern (and not premodern) understanding of energy as infinite *force* and profoundly limited available *resource.*

Thus we consider an ecological future not of Man or God but of the body and recalcitrant energy—not quantifiable, not refinable or concentrated in ways that allow for maximal inefficiency in the consumption of resources. Instead we posit an energy that traverses the body in ritual, in sacrifice, in its human-powered and unpredictable movement through the city—an energy that brings together a disunited society only to open it out to an unconditioned "night" that can offer no guarantees, a night in which the stars shine, heavenly bodies themselves radiating an energy too diffuse, too vast, too disordered, ever to be simply recovered.

In the coming years we will all become futurologists, whether we want this or not. We will be forced to think about energy: how its availability, at least in usable form, is constantly dwindling (amid the seemingly infinite quantities squandered in nature); and about religion: how it offers an alternative, a consolation, in modes of psychic or cultural satisfaction or warfare. We will then be forced to think about alternatives: energy alternatives, spiritual alternatives. We will be obliged to conserve and to recycle even as we recognize that recycling cannot be on the scale of cautious planning alone. This book is a small effort that tries to suggest that there are *other ways* of thinking about how we power our lives, with energy and with religion: these ways, these directions have been there all along. These other ways are not so much opposed to sustainability (as it is conventionally conceived) as they logically precede it and spin it off not as a goal but as an aftereffect. The energy of stars, always "lost" (nonrecoverable); the energy of our bodies in pain or ecstatic movement; the energy of sacrifice or religious orgy; the energy of art or recycled junk charged with insubordinate power—in and through all of this the "tendency to expend" may very well come to be recognized, and "experienced"—as we plunge into the deep recesses of human and inhuman activity. If in a future (and imminent) era of scarcity we rethink what it means to be happy—thereby recognizing that happiness is tied not to the mere consumption and disposal of materials, but to their wise use—we will perhaps also realize that happiness means something more, or other, than a meager conservation or a placid contentment grounded in a placid sociability.[10]

Now, in other words, is the time to start thinking about how we will spend and expend in the twenty-first century.

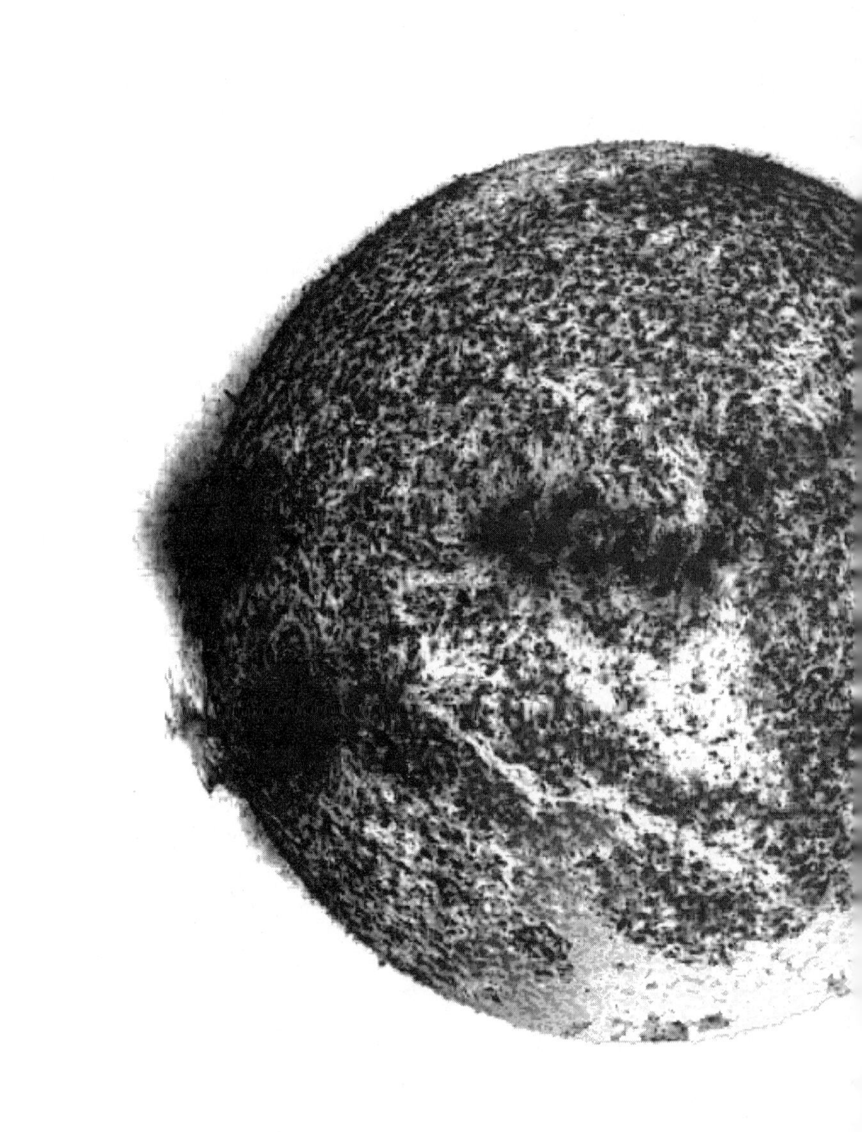

I. Rereading Bataille

1 Bruno, Sade, Bataille

Matter and Energy, Death and Generosity

Our story starts with matter. Is matter simply a neutral entity, fit only to be stockpiled, used, or discarded? Is it merely an inert object in relation to our active subjectivity, something we can appropriate to make our lives better? To guarantee us a period of comfort and satisfaction, our well-protected, nonnegotiable lifestyle? Is its *use* always to be conceived as leading to some higher good, something desirable beyond the present moment of satisfaction? Is matter to be conceived exclusively as an element of an object, a production that serves a human purpose?

Bataille's take on matter is crucial, and it is necessary to understand its genealogy. For Bataille, matter is infused with an energy (it "is" this energy) that is more than just the power to do work: profoundly excessive, Bataille's energy is inseparable from a subversive charge that both founds and undermines social stability. Bataille's matter is base, vile, and fundamentally resistant to any use, and only if we understand how Bataille arrives at this matter can we grasp Bataille's religion (of the death of God), his economics (which involve an energy that cannot simply be harnessed to do a job, and a matter that cannot simply be worked into stable and serviceable objects), and his ethics (a radical generosity based on the death of God—and Man).

In this chapter I will argue that Bataille's theory of matter and energy issues from a line of thought that can be traced through a certain alchemical, but also materialist, tradition. Giordano Bruno and the Marquis de Sade are crucial precursors for a number of reasons: they attempted to rethink the role of matter in creation, formulated a conception of God (Bruno) and ultimately of energy (Sade) that animates this matter, arrived at a dualist theory of creation through their emphasis on atomistic materialism, implied a notion of the death of God through their emphasis on an immanent (Bruno) or fictional (Sade) God, and made possible a theory of radical generosity (Bataille) out of a seemingly limitless monism (Bruno) and selfishness (Sade).

Bruno, Divine Materialism, and Religious Belief

Matter and energy were already closely linked in the Renaissance. Copernicus's great reversal, in which the sun came to reside at the center of the solar system, if not the universe, posed an enormous challenge to a theocentric model of creation: if heavy and sinful matter were no longer at the center (the earth), and ethereal matter, leading eventually to pure spirit, God, was not in the realm of the stars and beyond the stars, where then could God be? If, in other words, matter is everywhere, matter's relation to God must ultimately be reconceived. And with it, energy, because if mutable matter is omnipresent and active we can no longer think of a prime mover exerting force—the force of creation—in a one-way direction (anterior temporally, exterior spatially).

Giordano Bruno is an exemplary figure from the late sixteenth century who attempted to rethink the nature of matter in the context of Copernican heliocentrism. While Bruno has little use for the mathematics of Copernicus's calculations—he is a mystical philosopher, not a scientist—what is most striking is the way Bruno privileges matter. Matter is no longer to be identified exclusively with corruption, imperfection, decay; in fact it has a double nature, consisting of a matter that can be formed and a matter that is its substratum, but which is *formless*. Bruno writes in his pivotal dialogue *Cause, Principle, and Unity* (1998; first published in 1584):

> This natural matter is not perceptible, as is artificial matter, because nature's matter has absolutely no form, while the matter of art is something already formed by nature. Art can operate only on the surface of things already formed, like iron, wood, stone, wool and the like, but nature works, so to speak, from the center of its substratum, or matter, which is totally formless. (56–57)

Moreover, these two aspects of matter are perceptible through two different capacities: "artificial matter" is attained through a "cognitive principle" associated with the senses, whereas "natural matter" is perceived through reason (57)—no doubt through Bruno's philosophy itself.

"Natural matter" is the foundation of matter, the material used by nature to create individual forms of matter, which can then be transformed by art. Since this natural creation is dependent on matter, there seems little room in Bruno's doctrine for creation ex nihilo: the breakdown of one natural form leads to the recomposition of another. "Who does not see that corruption and generation derive from the same principle? Is not the end of the corrupt thing the beginning of the thing generated?" (99).

Nature is the principle, the agent, of this corruption and (re)generation; nature is therefore divided into active potency, "through which its substratum can operate," and passive potency, "through which it can exist, or receive, or have" (65). The formless substratum, in other words, exercises a given force, and the formed substratum receives the force and is moved: it takes on form. Most significant here is the fact that matter moves, is capable of moving, and that this movement is the action of nature. What a later age might call energy is a principle of nature, the driving force of all matter. Bruno reminds us that "active potency and passive potency are, in the end, one and the same thing" (65); just as are, no doubt, the natural matter and the matter of art (form-giving by the artist). All is one, inseparable, and infinite, as Bruno reminds us; there is no area from which matter or God (absolute possibility and absolute power) is absent.

Bruno takes great pains to argue that his model, unorthodox as it may be, is not an atheist one. If we step back from specific movements of matter, from specific forms, there is a unity, beyond time and location, which goes beyond any specific form or agency. The master metaphor for this unification is, not surprisingly, the sun: it describes a circle, but its existence cannot be separated from the totality of the circle, its position at all possible points.

> Thus, an indivisible [the circuit of the sun] is found to contain the divisible, and this is brought about not through any natural possibility, but through supernatural possibility—I mean, if one supposes this sun to be that which is in act all it can be. This absolute potency is not only what the sun can be, it is also what everything is and can be.... And so, what is elsewhere contrary and opposed is one and the same in [God], and every thing in him is the same. (68)

God, then, as "absolute potency" is the nonseparation of potency from act, the unity of material principle and formal principle (69). It should be recognized that he is also, however, the maintenance, the putting into action, of their opposition: its generation, not perhaps temporally (its creation out of nothing), since matter has always existed, but logically (the opening and maintenance of its possibility).

Bruno's model therefore entails a materialism in which God is present, indeed omnipresent, but where he is not simply separate from corruption or the transformations of matter. The demise of the hierarchical medieval worldview, with the earth at its center and God (distinct from corruptible matter) as prime mover, has led in Bruno's Copernicanism to a materialism where God is present as principle in all movement, all change; God is the atemporal presence that binds all of nature. All that was, is, and can be, God is; all creation, all mutation, all corruption, all death and life. All active and passive potency.

One notes the alchemical basis for this theory—the transmutation of substances in the principle of God's agency—but one also notes the extent to which God starts to seem superfluous in his own universe.[1] True, he is the universal principle, not only universal agency but also universal form; not only what can become, but what has and will become. He is the opening of the possibility of transformation of matter, from active to passive, from reason to the senses. But we must recall that earlier, in putting forward matter as a universal substrate, Bruno designated it as "natural matter" (56). It is not so much God who anthropomorphically intervenes, as nature that creates the specific forms of matter, which are then transformed by man. Iron, stone, wool, all these elements or compounds are clearly enough formed and re-formed not by the immediate, divine intervention of God's hand, but by processes that are, precisely, natural, and that depend on the breakdown of other forms of matter. The substratum with which nature works is perceptible only through reason, but the specific forms are clear enough, after all, to the naked eye. If wood rots, or iron rusts, some other compound or natural object ensues; it is not God who is causing this to happen—directly—but the movement of nature itself. God's might be the ultimate agency, but it is nature that is getting things done. And if we discounted God and simply studied the workings of nature—would it make all that much difference?[2]

This of course leads to a problem: if we need an "absolute potency" behind universal matter—the opening of the possibility of its formation, its opposition to itself—does it need to be the God of Christian religious doctrine? Can we speak of a universal potency, or universal principle of agency, that has nothing to do with the God of the Bible? Bruno, as we would say nowadays, didn't want to go there, despite the fact that his assertion of an infinite universe and an all-encompassing materialism were enough to get him burned at the stake by the Inquisition in 1600.[3] But the danger, even for a heretic, is clear enough: without God, with just matter, and along with it energy (the relentless movement of formation and deformation), there is little to serve as a foundation for ethics. Without man in God's image, what of the good? True, there is reason, which allows access to "natural matter"—but reason is a slender reed (hence Bruno's insistence on the presence of God in matter). To say that "corruption and generation derive from the same principle" only compounds the problem: if God is the principle of all that can be, the death and rebirth of any given animal will essentially be equivalent in his eyes. What counts, after all, is matter, its substratum, and its potency; from the perspective of matter there is little difference between the death and corruption—and regeneration in other creatures—of a person and, say, a dog. True, humans are the only rational animals, and therefore they alone would have access to a comprehension of God and his workings; but how can one separate their bodies from the reason that alone can comprehend natural matter?[4]

Bruno does not choose to recognize the horror of the infinite spaces that so tortured Pascal, the suspension between the infinitely large and the infinitely small. No doubt an alchemical operation could generate a human consciousness out of a seemingly indifferent nature, just as, for other alchemists, gold was derived from base metal. For Pascal, God alone could act to reassure man and give him a place between the two extremes; without God there would be merely the "disproportion of man." It seems, however, that by emphasizing God's identity with the potential of infinity, with the infinite proliferation of matter in opposition to itself, Bruno has underlined that very disproportion. Pascal's reassuring God is at least a moral principle of human existence and proportion; Bruno's God, in spite of the promises of alchemy, is ultimately the principle of natural, material multiplication (and subtraction). God is present in all matter, in every atom, but his presence is necessarily inseparable from the incessant transformations of those atoms, of that matter.[5] This is hardly all that reassuring: the vision of a rotting corpse, eaten by vermin, is an excellent example of corruption and generation, no doubt deriving from the same principle. And this is not at all a version of energy that can simply be put to work by Man. In a strictly material universe, Bruno's God can offer little consolation; where is heaven in Bruno's materialism? Could not God be replaced, for that reason, simply by a principle of matter, including an elementary matter that somehow generates the forms of perceptible matter, along with an inseparable energy that makes possible that generation? Replacing "agency" (which still implies some consciousness carrying out a plan) with "energy" (a measure of physical movement or transformation—"work") would be enough to bring about, so to speak, the death of God in the crudest, most literal sense: no more God. God would be inseparable from the very movement and finitude of matter itself, from the violent agitation of its atoms. And without God the primacy and integrity of the human, comfortably situated between the two infinities, would also be at risk. And no more Man as well; no more agency that could simply control energy, and for whom energy is always available, mastered.

The absence of God is already anticipated in Renaissance mysticism. One of Bruno's strongest influences, Ficino, affirmed the tradition of "negative theology." This line of thought, derived from readings of Pseudo-Dionysus and which certainly influenced Nicholas of Cusa in his work *De docta ignorantia,* held that since God is beyond language and beyond our senses, not to mention our intellectual capacities, we cannot hope to represent him linguistically or even to imagine him. No adjectives are adequate. As Frances Yates, in her masterwork *Giordano Bruno and the Hermetic Tradition* (1964), puts it:

> [Pseudo] Dionysus also sets forth a "negative way." There are no words for God in His actual essence; no names for Him as He really is; therefore He is at the least best defined by negatives, by a kind of darkness, by saying that He is not goodness, not beauty, not truth, meaning by this that He is nothing that we can understand by those names. (124)

If indeed this "hidden" God can be known only through (mystical) contemplation, one could doubt that one could say anything, produce any theological statement, about him at all. There are very fine gradations between a hidden, nonrepresentable God, an absent God about whom one speaks by not speaking, to a nonspeakable God, to, finally, a nonexistent God, a God whose very existence is denied in mystical visions. If there are no words for God, no place for him in any conceptual construct, how can we say that he "exists"? True, there is the evidence of "mystical" experience, and this certainly would have been enough for Bruno or for Nicholas of Cusa. To say that God cannot be represented, that his existence hollows out a kind of semiotic void, would seem to preclude any straightforward attribution of the mystical experience to God. Yes, there might be an emotional certainty of an experience of God, but if we cannot state it, represent it, how can we speak coherently of God's responsibility for the experience? Since we cannot speak of God or know him intellectually, what would prevent an assertion that the mystical event does not have to be tied to God at all? What if one argued that the experience is linked, quite to the contrary, to the nonexistence of God? That it is radically unconditioned?

If God is everywhere in all matter, if he is all that matter can be, he is nothing other than the potential for infinite mutability in space and time. And if God is to be identified with this mutability, then God is not unitary but the opening of duality, of opposition and change, the matrix out of which all force and all form arises, from which force and form is expended. He is the challenge to all stasis, all form, all labor (and its attendant goals), and all practice (including religious and mystical practice). Putting the sun at the center of the universe means putting matter at the center, matter with its endless transformations unattributable to a higher or more central authority. Without God as unitary Author of the universe, transcending all matter, there is a double or divided (divisible) matter (as Bruno has already argued), and therefore there will be a "principle" of matter that is always itself subject to movement, transformation, destruction, because it is not simply distinguishable from matter's duality (it cannot transcend it). A dualist philosophy recognizes matter charged with evil.[6] If we choose to call matter's principle God, then he will be both himself and contaminated by his radical absence, the impossibility of his own superiority and mastery. God, as the "origin" of the duality of matter, is not simply reducible to duality, but he is not separable from it either. He is the opening of the possibility of a dual matter without being simply double himself: he is logically prior to the incessant movement of the repetition and distinction of matter, its generation and decomposition through formlessness, into and out of forms. And we should not forget, he opens the distinction between sense (passive, formed matter) and reason (active, formless matter). Only in this way can God be seen to be radically unconditioned. Bruno himself never went this far in his argument, but perhaps he did not have to: following his lead, others drew out the heretical consequences of his positions and put him to death.

This God could literally be said not to exist. He is incessantly prior to existence, since he precedes the very distinction between the formed and the formless. Neither formed nor formless, divisible nor indivisible, God cannot be reduced to the duality of these terms; nevertheless, Bruno's God opens a field without hierarchy, a field of play and infinite movement (since he is not "above" matter or outside it, nor does he contribute to or crown a hierarchy of forms), *another* formlessness partaking of duality, of lateral transformations, but in no simple opposition to another term. This is his "absolute potency" and "absolute possibility," but also his death, his radical absence from himself as origin and end.

Sade: Atheism, Materialism, and Selfishness

The eighteenth century did without the theological niceties of Bruno's God. Bruno used Copernicus as a starting point, affirming not just the sun at the center of the solar system, but through the decentering of Man and God, an infinity of matter, spread consistently throughout the universe. Bruno's considerations, however, were not scientific in that, unlike Copernicus, Kepler, and Galileo, he was not concerned with observing and measuring phenomena and testing hypotheses. Bruno's God, inseparable from matter but not reducible to it, was conceived instead as a mystical, not a calculable, entity. But as we have seen, this God was also in a profound sense both within existence and on its other side; not simply reducible to being, he could be easily dismissed by later materialists for they held that, indeed, the one "thing" that existed, indisputably, was matter. With matter available to our senses, thus existing beyond any doubt, there was no need for God. Indeed all that comes to us is through the senses; mystical revelation and theological argument have no place in a modern, fully rational philosophy. The enlightenment might have seen Bruno's execution as yet another act of religious superstition, but it had no interest in following Bruno down the path of a seemingly improbable theological materialism.

Without God, we are left with matter alone; without grace, divine revelation, we are left with knowledge deriving exclusively from the senses. Following other, earlier eighteenth-century materialists—most notably Helvétius and d'Holbach—the Marquis de Sade attributes the greatest importance to sense experience: it is the source of all knowledge, hence all philosophy; it is the highest value; it is the source of all pleasure and morality,[7] and it is also, above all, the reason why we are all profoundly isolated.[8] For Sade, we are not only suspended between two infinities, but in a space in which *others* are infinitely distant. Without God, sensation alone is our raison d'être; it is also the measure of our radical distance from, and superiority to, others. As Dolmancé, the libertine hero of Sade's *La Philosophie dans le boudoir,* puts it:

> There is no comparison between what others feel, and what we feel; the strongest dose of pain felt by others must assuredly be nothing for us, and the slightest tickling of pleasure felt by us touches us; thus we must, at whatever price, prefer this slight tickling which delights us to this immense quantity of others' misfortunes. (1968, 172)

Radical materialism without God, in other words, leads to a cult not only of sense experience, but of the self. Given the isolation of the self, all possibility of fraternity is lost. Instead, morality turns around whatever stimulates my senses in an agreeable way; and if, "as is often the case" (173), the "singularity" of my organs is such that my pleasure depends on the pain of others—well, in that case, I am morally obliged to delight myself and cause pain to others. In other words, it is Nature "herself" ("she" is always personified by Sade) who is demanding this cruelty. What communication there is between beings lies in this inflicting, or reception, of pain—to the point of death.

Since it is Nature who has determined our pleasure—by constituting our bodies in such a way that stimulus is both enlightening and agreeable—it is Nature who determines our morality. "She" not only determines that the infinite distance of others guarantees that their pain is my pleasure, but she also determines that my crimes—my murders—are useful to her larger order.

Nature, in other words, needs the death of her creatures; death is not definitive but a point of transition in a larger movement of life. As the author of Dolmancé's favorite pamphlet, "Français, encore un effort si vous voulez être révolutionnaire" ("Frenchman, One More Effort If You Would Be Revolutionaries"), puts it:

> Now, if her [Nature's] acts of destruction are so useful that she absolutely cannot do without them, and if she cannot arrive at her creations without drawing upon this mass of destruction prepared for her by death, from then on the idea of annihilation that we attach to death will therefore no longer be real; no longer will any annihilation be perceived; what we call the end of the living animal will not be a real end, *but a simple transmutation, whose basis is perpetual motion,* the true essence of matter and what all the modern philosophers consider to be one of the first laws. Death, according to these irrefutable principles, is therefore nothing more than a change of form. (1968, 258; italics added)

One can see in this quote, quite clearly, Sade's revision of the Renaissance mystical tradition represented by Bruno. Nature is matter, and matter is in perpetual movement; the death of one creature and the resulting birth of another is nothing other than a "transmutation," the alchemical transformation of one substance into another.[9] As with Bruno, the movement of matter is immanent (matter and movement being interchangeable: matter *is* movement); the difference in Sade is that mystical enlightenment associated with alchemy is replaced with extreme sensual pleasure. And, perhaps consistent with this, rather than a movement upward—from a base metal, say, to a noble one—we see a lateral movement, the simple transformation of one animal into another, neither of which is privileged (by nature): a human into a fly, for example. We could even argue that Bruno's God, the absolute potency and possibility of matter, is replaced in Sade by energy: energy is contained in all bodies, *is* all bodies, and is liberated in them (and liberates them) through acts of violence. It is this inexhaustible energy, ultimately, that is inseparable from the constant violent movement of atoms that is Sadean transmutation, destruction, and rebirth. Energy is the force in Sade that subtends the constant transformation of forms, of lives.

Not only is death not a definitive end; according to the pamphlet's author, nature once again dictates human morality (or immorality) in that the extreme pleasure of violence—of murder—serves to contribute to nature's processes: "the action you commit [murder], by varying the forms of her [nature's] different works, is advantageous to her" (260). Moreover—and this is crucial—by committing murder, humans *speed up* the natural process, in a sense making it even more natural; the true alchemy of man is a doubling of nature's alchemy, but a doubling that intensifies the violence of the process. "Little animals," writes Sade's fictional pamphleteer, "are formed at the instant the larger animal breathes its last" (259).

This recycling, then, is quite different from what we today might associate with that term. "Natural" recycling for Sade entails not primarily a parsimonious, calm, and virtuous reuse of available material; instead, it entails the alchemical manifestation and intensification of a violent energy inherent in nature and in matter. The basis of matter is, as we are reminded, "perpetual motion"; this motion is only revealed through the highest moral/immoral act to which humans can aspire: murder. It is in the crucible *(creuset)* of Nature that matter comes into its own as matter.[10] In the incessant act of transmutation, matter is most what it is: sheer movement and the stimulation of the senses (the reward for bringing about transmutation) that is inseparable from it. Here again, of course, Sade reveals himself an eighteenth-century philosopher: for his contemporaries as well (d'Holbach, La Mettrie, Helvétius), matter was the ceaseless agitation of atoms, God was irrelevant or absent, and morality was a function of what was pleasurable.[11] Sade intensifies this gambit by demonstrating how this essentially formless matter—formless because it has no inherent form, but only the inherent energy of "transmutation"—is inseparable from the necessary violence that nature "desires" and dictates between creatures. Between, in other words, humans, between sexual predators and their victims. Matter is not only fundamentally active; its activity stimulates the senses of the creatures that embody matter and, thus stimulated, they kill. Their acts of destruction (and in the case of humans, murders) inevitably close the cycle by casting more matter into the "crucible," freeing more energy and leading to more "transmutation." This destruction-creation, moreover, is desired by Nature and rewarded by her: it seems that the Sadean libertine finds murder, even the destruction of whole populations and civilizations, pleasurable precisely because Nature has willed it.

Recycling, then, is not primarily practical but orgiastic, the violent, ecstatic, and agonizing production, perversion, and destruction of forms. Its practicality, if it has any, is an aftereffect of this movement. With the incessant reuse of matter comes not stability but the unending loss, in agony and delirium, of coherent lives, meanings, laws, and even societies, accompanied by the extreme pleasure of those who carry out the destruction (and inevitable reuse).[12]

For Sade, matter is dual, as it was for Bruno: the endless transmutation of matter into different beings and forms is inseparable from the expenditure of the coherency of those forms and the reined-in (usable) energy associated with their maintenance. There are always new forms, those forms are unstable, and their constant change is accompanied by an infinite expenditure of energy. Matter is constant movement, constant disruption, an endless burn-off of energy; it is also the momentary forms that this energy establishes, maintains, then destroys. Infinitely expended, the energy of matter in Sade therefore plays the same role that God plays in Bruno: it opens the distinction between formed and formless matter and is the principle, and the agent, of their interaction. Energy is the "absolute power."

The fictional pamphleteer of *La Philosophie dans le boudoir,* and Dolmancé along with him or her, comes off as a hideous rewriter of Rousseau:

Nature now is not the benign state to which we aspire but, it seems, is instead the violent matrix (with all the feminine implications of that word) that cruelly takes away what it so capriciously gives. But perhaps predictably, Sade has only marginal control over the metaphors and personifications he proffers: Nature, as a seductive, despotic, and murderous "she" (a Sadean heroine, in other words, another Juliette), "herself" enjoys the glories but also runs the risks of any character in Sade's pantheon. She also can, in other words, be defied, defiled, murdered. She and her panoply of forms (and lives) may not be prior to energy after all. The very anthropomorphized agent by and through whom coherent forms come into being is at risk. This is a point raised by Maurice Blanchot when he notes the paradox of the criminal: "If crime is the spirit of nature, there is no crime against nature and, consequently, no crime is possible" (1963, 41).

Sade confronts this scandalous fact by finally having his heroine Juliette deny and insult Nature just as she earlier denied and insulted God. In both cases, a figure sets against man a law that is inviolable: goodness in the case of God, crime itself in the case of Nature. Even the coherent "natural" law by which forms are given and orgiastically transformed is liable to destruction. But how can one break the law of crime?

Blanchot argues that there is a kind of all-encompassing negation embraced by the "sovereign" man. As he points out, "The integral man, who entirely affirms himself, is also completely destroyed" (44). Destroyed because he is no longer even human; by committing his crimes, he does not so much act on his desires as serve as a kind of concentrating point for an impersonal energy, one that has been repressed, as Blanchot puts it, throughout "seventeen centuries of cowardice" (44). Blanchot's main point is that this energy is what remains (indeed retains its primacy, its logical priority) after every certainty—God, Man, Nature—has been attacked, sullied, and destroyed.

The radically apathetic Sadean hero or heroine refuses all emotion, all humanity—all pity, gratitude, love—that would serve to weaken and debilitate. The hero/heroine literally becomes "insensible" in the moment of greatest destructive energy; from this insensibility there is then a passage to an all-powerful state, and no doubt back again, when the sovereign figure shows a sign of weakness—of dependence, of humanity—and is in turn devoured or destroyed by another.

The greatest energy, then, supplanting Nature herself, entails a radical selfishness that finally is not even selfish: it is a kind of impersonal power that concentrates its forces in order to heighten the effect of energy itself, intensifying and hastening the process of destruction. The Sadean hero is a mere conduit of this energy, him or herself "dead," a kind of zombie through which energy applies itself. One could clearly ask—though Blanchot never does—whether this energy is in turn not merely another law that would have to be broken. But energy would seem to have a special status: it is not so much an entity or a law, and certainly not the power to do useful work, but rather the absolute principle and power out of which laws and that which they oppose arise and by which they are destroyed, the principle and motivation of formed and formless matter. Hence the Sadean sovereign, endlessly affirming death: not literal death, of the sort undergone by the weak, virtuous victims (weak because they are virtuous), but rather death in energy, the radical destruction of anything that could remotely be considered human. The total selfishness of the Sadean hero results not only in apathy but in a paradoxical and hideous generosity: the giving, the sacrifice, of energy with no concern for the maintenance of the self, of the human, of one's own humanity, which is now "dead." Sade, however, while affirming death, never affirms the generosity inherent in his model of self-destruction.

We have moved, then, from matter in Nature to something that is more fundamental than Nature's matter, or that characterizes matter beyond matter itself: energy. Energized matter, after all, risks stability, a given form, even if that form represents only the briefest existence. Nature's laws are just that—weak laws to be broken. Energy, on the other hand, is movement itself, or rather the concentration of movement that can never be stopped, never be altered. Like God in Bruno, energy in Sade opens the possibility of the opposition of formless and formed matter, since it is responsible for the transformations of matter (from one form to another, via formlessness); it is ultimate impersonal agency, unmasterable because it is both invested in but not reducible to the opposing terms in and through which everything is created and destroyed. It "is" formless matter, the matter of transmutation, just as it "is" the disruption of the formed matter that has been thrown up out of base matter. In energy the Sadean sovereign attains the destruction of everything at the cost of a death more profound than simple death.

Needless to say, it is impossible for Sade to derive any kind of workable morality from his dystopian realm of monsters. As Sade sees it, the law, indeed any law, attempts to restrict arbitrarily and unjustly the fundamental and natural movements of energy and pleasure. God for Sade is the ultimate instance of an innocence that can only be preserved through a repressive and necessarily unjust "cold" law. At best Sade can offer, perhaps as a parody, a society in which God is absent and laws are light, virtually nonexistent (law as the abolition of law)—so that murder is only "punished," if that is the word, through revenge.[13] Precisely the situation, one might object, that the law was meant to alleviate in the first place.

But we take away from Sade's cult of the inhuman a vision of sovereign energy that in a sense transcends everyday selfishness, since its highest point is extreme apathy; can a sovereign, steeped in murderous disinterest, dependent on nothing and no one, still be said to be selfish? Does a sovereign self-sacrifice, in which even the self is immolated to maximize energy, entail an extreme self-centeredness, or its opposite? Bataille, in fact, will imagine a generosity derived from Sadean apathy, running *against* the Sadean cult of individualized nonhuman energy.

In any event, by pushing destruction so far that the sovereign's very being is annulled, Sade has reaffirmed the "crucible," perhaps not of nature, but certainly of energy. A crucible in which matter is no longer the solid, stolid thing that virtuously allows itself to be used, put to some higher purpose in the tool, but is instead the play of ferocious, incessant agitation—untouchable and repugnant in its decomposition (and recomposition). A crucible in which energy does not just do work, but plays as the force of the expenditure of forms. The movement and loss and replenishment of energy is inseparable from the violent destruction and recombination of forms that is always accompanied by some reaction, some pleasure: all we can value is inseparable from physical pleasure, or the ultimate renunciation of pleasure that goes by way of pleasure (the itinerary of the Sadean hero).

And yet one could argue (*pace* Blanchot) that something human remains. For the Sadean monster at his or her highest (or lowest) moment is still *conscious* of this complete renunciation of feeling, satisfaction, pleasure. If the sovereign melds with the energy flow, concentrates it, he or she nevertheless is still aware of the process, is still guiding it on some level. This paradoxical awareness—of that which is the experience of death and the formless continuity of all (necessarily dying) beings in sheer movement, as energy expenditure—would seem to lead to a human grasp of something utterly beyond anything one could characterize as human identity, morality, or society. Sade in this sense needs the human, just as he needs

God: he needs the directing consciousness of the selfish master, just as he needs the weakness of the victim, of the isolated human consciousness, and along with the human victim, the ultimate victim, God himself. But he refuses to admit it: he (and his characters) cling to an eighteenth-century atheism, the simple refusal of God, his "refutation." And yet Dolmancé tells us he is frustrated: without the blaspheming and murder of God, the ultimate virtuous victim, there can be no orgasm:

> One of my greatest pleasures is to curse God when I have a hard-on. It seems then that my mind, a thousand times more excited *[exalté]*, abhors and has contempt for that disgusting chimera to an even greater degree; I would like to find a way either to denounce him better, or to insult him more; and when my accursed reflections lead me to the conviction of the nothingness of this disgusting object of my hatred, I become irritated and I would like to reestablish the phantom, so that my rage could at least be directed against something. Imitate, me, dear lady, and you will see the intensification that such discourses will infallibly provide your senses. (Sade 1968, 103–4)

Sade in fact needs a dead or dying God, a God who endlessly gives himself to be murdered (which is not the same as a God who simply does not exist). God, in other words, as the internal limit that throws up meaning, law, sense, virtue, only to accede to his own destruction, his ultimate victimization, at the moment he recognizes, or submits to, the finitude of his own meaning. A dead God who incessantly sacrifices himself. Sade was no doubt looking for such a figure, his "philosophy" needed one, but he could not believe in it. Just as Bruno believed too much in a God who, by not transcending matter, managed to throw into doubt his very own existence, so Sade did not believe enough, and in the process only reaffirmed a profoundly selfish, indeed solipsistic, humanity (the individual power of Sadean heroes). Sade refused to believe that one could affirm, if not simply "believe in," a dead God, a God who had given himself to the point of sacrificing his own divinity. By refusing this God, by clinging to a too-simple atheism, Sade denied himself the very possibility of writing the death of God in society. His fictions were condemned to be fictions; they could not pass on to the emptying, the giving, the sacrifice, of the sense of the divine, of the human (since the divine is nothing more than an intensification of the integrity—the weakness—of the human). By retaining a simplistic atheism, a simple refusal to accord the virulence of God's death, Sade ended up retaining, on a number of strata in his texts, the human. If God simply does not exist, the human steps in for him: where there is absence of God, there is the plenitude and power of Man, the presiding consciousness of the sovereign. By not recognizing the profound virulence of the death of God, Sade fails to grasp the implications of the death of Man. In Sade, men and sovereigns may die, but in their death they nevertheless act as the focal point of a profoundly selfish energy, socially inscribed in a powerful and all-pervasive fiction. The Sadean sovereign, embodying immorality and truth, becomes ever more powerful, directing ever more energy, as the victim, the embodiment of lies and virtue (the lie of virtue), becomes weaker, a complete deficit of energy. (The ultimate deficit, the ultimate zero—and the ultimate lie [or fiction] is God, believed-in as dead.) The energy wielded by the sovereign might come to "replace" God, but in so doing it is inevitably reappropriated in a fiction of the supreme and selfish power.

It was Georges Bataille who would continue along the path marked by Sade in his attempt at writing the dead God and his link to a virulent matter and an uncontainable energy, in the context of generosity and the death of Man.

Bataille, Base Matter, and Generosity

Georges Bataille's writings from the early 1930s make clear the fact that he was quite explicitly engaged in a rewriting of Sade—at that point, in fact, Sade's influence on Bataille was much more important than that of Hegel or Nietzsche. "The Use Value of D. A. F. de Sade," a polemical text dating from the early 1930s, contains a number of themes that Bataille will go on to develop in the immediate prewar years: most notably, that there is a "heterogeneous" matter whose very virulence prevents its inclusion in any "closed economy" of use, practicality, or recuperation. This conception of matter is clearly derived, in part, from the anthropological theory of Durkheim and Mauss. For these authors, the (left-hand) sacred or taboo object is not simply something to be accorded a special status; it is, on the contrary, infused with a certain contagious power, analogous to an electrically charged piece of metal that can transfer its energy—with a spark and a shock.[14] For the French anthropologists, these taboo objects are constitutive of the community: their "electrical" energy is specific to a group, or clan, that includes them in their rites. Ordinarily untouchable, dangerously contaminated, these objects (or foods) nevertheless are handled (or eaten) at specific times, in specific rituals, and thereby confer their energy to the group to which they are sacred. Durkheim's anthropological project, in effect, is the effort to isolate the rational thinking behind this irrational perception of the energy of the taboo object, thereby transforming a left-hand sacred of untouchability and revulsion into the right-hand sacred of social coherence and the elevation of a higher meaning. The group's cohesive energy is transmitted, or communicated, through its collective contact with the taboo object. Then real energy, Durkheim maintained, is that of the collectivity, when it comes together, experiencing the strength of its solidarity and recharging its emotional batteries for future tasks and struggles. In mystified, irrational form, this energy is still perceived as an autonomous, sacred force: "mana."[15]

Now in Bataille this force, while explicitly associated with contemporary anthropological theory (as in "The Notion of Expenditure"), is also associated with Sade. Not that Bataille is attempting to argue that Sade was an anthropologist *avant la lettre;* the stakes are higher than that. For while the French anthropologists are attempting to uncover the rational bases of all societies, "primitive" and modern, and the rational bases of social energy, Bataille is attempting, through Sade, to do the very opposite. Rather than deriving a right-hand sacred from the left, Bataille hopes to liberate the left from the illusion of the right. The "science" of heterology, for Bataille, is the science of the force, the energy, that *cannot* be appropriated by science (or by an organized lawful society, organized under the heading of science or religion): an energy that cannot be given meaning by a higher principle that is itself useless, dead. Rational science, mathematics, physics, and, certainly, the social sciences reduce all questions, all phenomena, to quantifiable data that can be inserted into a "closed" system, one that solves all problems adequately. This is the ideal of science: even if problems cannot all be solved at once, the eventual goal is to leave nothing unanswered, to leave no mysteries. The sacred violence of matter is reduced to a timeless formula. But inevitably there are things that cannot be assimilated to any scientific understanding because by definition they are "heterogeneous": they are the terms that underlie the scientific method, Bataille argues (in a proto-deconstructive mode), but whose exclusion is necessary for that system to constitute itself in its homogeneity, its coherence. Bataille writes in "The Psychological Structure of Fascism": "Heterogeneous reality is that of a force or shock. It is presented as a charge, as a value, passing from one object to another in a more or less arbitrary way" (*OC,* 1: 347; *VE,* 145).

Bataille stresses that this expulsion of (by definition) heterogeneous (inassimilable) terms is functionally identical to the excretion of shit by the body: "As soon as the effort of rational comprehension leads to contradiction, the practice of intellectual scatology requires *[commande]* the excretion *[déjection]* of inassimilable elements" (*OC,* 2: 64; *VE,* 99).

Thus Bataille posits a "science" of what cannot be scientifically known or of what rigorously resists scientific knowledge: not shit as mere matter, but shit as base, charged matter that defies assimilation into any scientific or classificatory grid. Moreover, there is no distinction possible between charged, disgusting matter and revolting intellectual processes that lead to no higher synthesis (e.g., gnosticism). Thus Bataille's own project comes to be associated with the heterogeneous matter and horrifying forms that he studies, since it can lead to no satisfying "higher" conclusions or affirmations.

This heterogeneous, or charged, matter circulates, spreading its violence in a way that was first identified by Sade in a description of a coprophagic orgy: "Verneuil has somebody shit, he eats the turd and demands that someone eat his. She who has eaten his shit pukes; he laps up her vomit" (cited in Bataille, *OC,* 2: 59; *VE,* 95). While for Sade, perhaps, this description of shit is merely a titillating detail supplementing others, for Bataille it is central, because it posits a matter that is repulsive and untouchable not in its being but in its circulation (its appropriation/rejection). This is matter, to use Sade's term, in the act of transmutation, but here it is not a question of the (overly) rapid transformation of, say, a human body into worms or flies. Rather there is a process of exchange in which the matter is excreted, appropriated, excreted. At each point of excretion/appropriation—the two actions are reciprocal, inseparable—at each point of recycling, in other words, there is a charge of pleasure or nausea on the part of a desiring, or dying, subject.

The intellect, then—science—excretes and devours, and this process is no different from what Sade's "libertines" do. The process of "scientific heterology" entails nothing other than the rigorous study of all forms of rejection of inassimilable elements, be they "intellectual" (monstrous theories, grotesque art works, hideous poetry) or "physical" (shit, flies, big toes, spiders, spit). The products of such study—Bataille's own writing—will similarly be heterogeneous: disgusting and unassimilable.

Rather than discovering the rational bases of collective enthusiasm—Durkheim's goal—Bataille seeks to "study" the very matter whose unjustifiable, a priori exclusion makes possible the coherence of rigorous, hierarchical systems of classification and thought. There is nothing inherently heterogeneous—repulsive, nonappropriable—in shit or in anything else. It is the relation of that element, that object, to a system in which it cannot be given a stable position that makes it "rotten." Its excluded rottenness is necessary to the coherence of the system. As Bataille puts it: "rottenness is not just a middle term between the ear and the wheat. The inability to envisage, in this case, rottenness as an end in itself is the result not exactly of the human point of view but of the specifically intellectual point of view (to the extent that this point of view is in practice subordinated to a process of appropriation)" (*OC,* 2: 65; *VE,* 99).

Rottenness is an end in itself: it does not "go anywhere," but instead wanders in the labyrinth, throwing off energy, disorder. Heterological science studies it, but not as a means to something else. Bataille's science attains not a higher knowledge but a failure point of knowledge (just as to eat is always to vomit): "a burst of laughter is the only imaginable, definitively terminal, outlet—and not the means—of philosophical speculation" (*OC,* 2: 64; *VE,* 99).

A matter that is an end in itself, leading nowhere; but this matter is also always prior, an unformed, or definitively formless, "stuff," *mana,* which circulates through entities and communities, which opens their possibility, but which is never reducible to them. It is the charged, energy-laden material that draws together, and sets apart, Verneuil and his victims; and in another context, it is the junk that ties together people in primitive religions at the moment they expel it—when "it," for example, is the mutilated corpse of a sacrificial victim. It is the energy of their violent communication. "What [formless] designates has no rights in any sense and gets itself squashed everywhere, like a spider or an earthworm" (*OC,* 1: 217; *VE,* 31).

Thus for Bataille this sacred, charged, formless matter entails a religious experience, if one could use that word, but it is an "experience" that is not to be associated with established, Christian religions. (Bataille writes, "WE ARE FEROCIOUSLY RELIGIOUS, and, to the extent that our existence is the condemnation of everything that is recognized today, an inner exigency demands that we be equally imperious" [*OC,* 1: 443; *VE,* 179].) Christianity proposes a homogeneous sacred, one that attempts to assimilate all evil, all rottenness, to its own system: evil is only the absence of God, it is not a positive entity, and one can work one's way out of it on the way to a sacred belonging (Dante). The Christian sacred is ineffable, pure, holy, elevated above the rotten matter it must distance itself from in order to constitute itself. But Bataille notes the existence of early, now banned and cursed religions that maintained a cult of "base matter": gnosticism.

> In practice it is possible to give as a *leitmotiv* of Gnosticism the conception of matter as an *active* principle having an eternal, autonomous existence, which is that of darkness (which would not be the absence of light but the monstrous archons revealed by that absence), that of evil (which would not be the absence of good, but a creative action). (*OC, 1*: 223; *VE,* 47; emphasis Bataille's)

Religion returns as a cult of materialism; not useful matter, to be sure, but "formless" matter that communicates its energy ("creative action") between members of a cult. An elevated, pure God is doubled by a dead God or a multiple cursed divinity (such as the archons of gnosticism). And along with this matter there is an energy, given the Sadean context of Bataille's writings of this period, that is inseparable from a death drive. In some ways this "religion" is more rigorously atheist than was Sade's; Sade, as is clear enough in *La Philosophie dans le boudoir, needed* God: the Sadean orgasm could only reach its peak intensity when the libertine was engaged in blasphemy. Sade was in effect already developing a model of sacrifice *avant la lettre:* God's sacrifice was integral to the transgressive act. The sacrifice of God and the production of "sacred matter" (the expelled, murdered, formless mess left over after the orgy) were inseparable. Sade's dead God and his fiction are coterminous.

Bataille's religion, then, must take up from where Sade's necessarily blasphemous nonreligion ends. In a sense Bataille must be conscious of the religious double bind in which Sade finds himself, the necessity of positing God only to blaspheme and murder him. But where Sade, locked in his fiction, hits a dead end, with Dolmancé regretting that he cannot believe in the God he blasphemes, Bataille affirms the virulence of the dead God erupting in and against society. This is the practice, if one can call it that, of Bataille's heterological science, a science that ends not in self-satisfied knowledge but in the horror of a revolting sacrifice. God exists only in his death, his expulsion, his violation. It seems that Bruno's immanent, material God has come back in Bataille, but with his status as uncontrolled eruption of energy now made explicit: endlessly opening the violent duality of matter, God does not "exist," but he incessantly dies, since the matter he inaugurates is the very destruction of any coherent principle—and God is the ultimate principle. If laughter is the terminal point of scientific activity, then God is the terminal point of laughter: "When God realized he was an atheist he died laughing." Bataille's dead God is not mere fiction, but myth and tragedy (tragedy being the highest myth, that of the death of the king, the leader, the most high).[16]

It's here that Bataille's atheism intersects with Marx's; in "The Use Value of D. A. F. de Sade" Bataille posits a proletarian revolution that does not only involve the guarantee of workers' rights, or workers' comfort. Rather, it is based on the principle that the proletariat is, in the modern world, the privileged vehicle of heterogeneous matter; the workers alone are still in contact with the profound, base experiences (orgiastic sexuality, mindless spending, death) that the bourgeoisie has long since abandoned. "In the final analysis, it is clear that a worker works in order to obtain the violent pleasures of coitus (in other words, he accumulates in order to spend)" (*OC,* 2: 65; *VE,* 99)

This notion of a revolutionary expenditure (or, in the terms of the Sade essay, defecation), linked to an abject class, stayed with Bataille throughout his life.[17] I would stress here the extent to which Bataille has appropriated Sade's "politics," *quite literally.* Revolution, as it is presented in "The Use Value," is nothing other than the Sadean republic we recall from *La Philosophie dans le boudoir:* a republic of "soft" (nonexistent) laws that will encourage crime. Whether Bataille wants to go all the way, so to speak, and celebrate murder, along with Sade, is not clear; what is clear is that, like Sade, Bataille's postrevolutionary phase will entail the fundamental repudiation of "harsh" law—i.e., law in general, and the affirmation of a general transgression of law. It will be the societal equivalent of intellectual scatology: the point at which law is affirmed only to be blasphemed. Bataille says this of revolutionary activity:

> But the post-revolutionary phase implies the necessity of a break *[scission]* between the political and economic organization on the one hand and on the other an anti-religious and asocial organization having as its goal orgiastic participation in various forms of destruction, in other words the collective satisfaction of needs that correspond to the

> need of provoking the violent excitation that results from the expulsion of heterogeneous elements. (*OC,* 2: 68; *VE,* 101)

We are very close here to the ideal form of government presented in Sade's *Philosophie dans le boudoir,* an atheist republic in which prostitution is generalized—any citizen is obliged to engage in or submit to sexual perversion with any other, at his or her demand—and the "cold" laws of the rational, brutal, and *legal* Terror are replaced by self-effacing "hot" laws in which crime—up to and including murder—is tolerated, indeed encouraged.[18]

The problem, of course, is that, like Sade's God, and like his characters too, Bataille's hot republic will end up being profoundly fictional—a mental construct erected only in order to be violated, a wild story that disappears as quickly as it comes to life.

To say this is not to dismiss Bataille, but merely to recognize the difficulties associated with such a literal appropriation of Sade. The paradox of Bataille, as we've already seen, is contained in the phrase "Heterological Science." Bataille's science would be as rigorous as official, "homogeneous" science, while at the same time recognizing its defeat, its downfall as an accurate depiction of the world—that is, of stable, formed, not-quite-dead matter. Like God, like laws, science would be necessary above all to pose or open the terms of its own violation, its own extinction. A mythical science, in other words, or a rigorous science devoured by its own recognition of its dependence on myth: myth, like laughter, is what would typically be taken as a middle term, now established as definitive end. Myth is fiction that is in a sense more true than truth, a knowledge higher—and lower—than the highest knowledge of philosophy.[19]

Key to this mythical status of Bataille's argument is the profound Sadean selfishness I have already mentioned. In some sense this deep solipsism, in which only my pleasure counts, is Sade's greatest fiction, the only one, really, that allows the elaboration of all his other multitudinous narrations. If the self is completely alone, if the tiniest bit of my pleasure is more significant than the greatest agony you can experience, then you (the reader, any "other" who is by definition less powerful, less real) are already only a fiction, to be called out of nothingness, fucked, destroyed, and recreated at will. You can only be a simulacrum in a simulacral scenario. But then again so am I, since I need your simulation—your fictional status—for my own pleasure. My own pleasure is just another term in a network of fictions. This raises a question that is at the heart of Sade's project: how "seriously" are we meant to take his criminal revolution?

There is no easy answer to this question. It is easy enough to read *La Philosophie dans le boudoir* as a kind of enormous *Modest Proposal,* a grotesque parody that makes us aware principally of the extent to which crime had come to reign supreme in revolutionary France (the book was published in 1795). But it is just as easy to take it as a "serious" proposal, as Caroline Warman does in her excellent book (2002): Sade, in her view, is very carefully drawing the logical conclusions to the eighteenth-century theories of materialism of d'Holbach, La Mettrie, and Helvétius. The same problem appears in the case of Bataille, but there, in a sense, it is more explicit, and it receives a more consistent elaboration. Bataille recognizes, as we've seen, the paradoxical status of a science of heterology—the fact that the science is lost, devoured by the radically mythical status of material that defies the very possibility of sober analysis, quantification, and systematization. He recognizes, in other words, the myth at the heart of heterology. Nevertheless, he presents an analysis that would take Sade "seriously" as a precursor to Marx: a writer who enables us to situate the working class in a social model that implies not merely production but the kind of delirious consumption previously associated only with tribal chieftains. For that is proletarian liberation for Bataille: giving the workers not equal access to consumable goods, not the serious payback for their serious labor, but equal access to orgiastic and ludicrous destruction; to crime, in other words.[20] Bataille rewrites Sade, liberating crime, in a sense, taking it away from the aristocrats, kings, and popes who are the greatest criminals in Sade's fictions, and awarding it to the working classes. He does what he knows to be "impossible": he takes Sade's radical fictionality seriously, rewriting it as virulent myth, just as he takes the death of God seriously, pushing Sade's naive and cynical atheism to a level in which divine mortality is the supreme instance of the expenditure of energy, the ultimate blowout of the system of law, knowledge, and society. For Sade the death of God is ultimately a fiction, a masturbatory scenario along with all the others, whipped out of nothing because God "does not exist"; for Bataille, the death of God is a myth because God's death is not a simple plot device, but rather an instance of a cataclysmic recognition of God's necessity as well as his fall, his central and overwhelming "presence" as heterogeneous force of the sacred in society—precisely in his radical absence. Myth, like tragedy for Bataille, is a construct that reveals this absence, this death, in stories, in figures (like the archons of gnosticism), in the charge of sacred objects. Myth is therefore what conveys best the "formlessness" of a knowledge that eludes the vain practicality and anthropocentrism of the natural sciences.[21]

Bataille's science of heterology, in other words, interjects itself against Sade's all-encompassing fiction. For Bataille still retains a model of generosity, a perverse one, to be sure, but a model of giving that recognizes the other in a way completely foreign to Sade's antisocial and cartoonish society. By stressing generosity over a monstrous selfishness, Bataille opens the way for the myth of the death of God and the recognition of its consequences to be implemented in society.

If, in fact, as we know from Freud (and Bataille is happy to make use of the Freudian model of anality), defecation is a kind of gift, then cursed matter is caught up in a relation of generosity even more powerful than Sade's endless relation of selfishness and murder. It is not so much the model of violence that escapes fictionality, but that of generosity. It is through generosity that Bataille can elaborate a social model that recognizes excessive energy—an energy that cannot be reduced to useful and significant tasks—at the heart of all social activity. This social model, or sociology, is itself mythical, but mythical to the precise extent that it takes into account a reality that can only be excluded (as charged, heterogeneous) by conventional (social) science.[22] Generosity entails God's gift of himself as dead victim (Jesus doubting, profoundly, as he dies on the cross) and Man's gift of himself through the sacrifice of the coherent consciousness, sense of purpose, and higher (scientific, religious) knowledge that he (mistakenly) assigned himself. The gift of Man as the death of Man, in other words.

Bataille's thinking on generosity clearly evolved over the years. While the "Use Value of D. A. F. de Sade" essay, written around 1930, seems to value a fairly straightforward Sadean utopia/dystopia, a postwar essay on Sade, "De Sade's Sovereign Man," first published in 1949 and reprinted in *Erotism* (1957), makes clear that Bataille intends to appropriate Sadean cruelty by rethinking it as generosity.

In "De Sade's Sovereign Man," Bataille first broaches the issue of generosity in the context of the radical apathy analyzed by Blanchot in his essay on Sade. As we have seen, Sadean subjectivity is nothing more than an intensification to the point of explosion (or implosion) of violent energy.

Bataille shifts the emphasis of Blanchot's version of Sade slightly by emphasizing a link between this self-destruction and generosity. Bataille stresses in his first Sade essay, in *Erotism,* the fact that "denying others in the end becomes denying oneself" (*OC,* 10: 174; *E,* 175). Bataille, however, asks: "What can be more disturbing than the prospect of selfishness becoming the will to perish in the furnace lit by selfishness?" (*OC,* 10: 174; *E,* 175). The extreme selfishness of Sade's characters thus turns against selfishness: when extreme pleasure is pushed to the limit, the sheer energy of destructiveness threatens the stability of all selfish subjectivity.

Bataille goes on to illustrate this self-destruction with a reference to a character, Amélie, who appears very briefly in Sade's *Juliette.* Amélie, though quickly sketched, is striking because she offers herself as a crime victim to one of Sade's most sinister villains, Borchamps. Bataille cites a passage in which Amélie addresses Borchamps:

> I love your ferocity; swear to me that one day I shall also be your victim. Since I was 15 my imagination has been fired only at the thought of dying a victim of the cruel passions of a libertine. Not that I wish to die tomorrow—my extravagant fantasies do not go that far; but that is the only way I want to die; to have my death the result of a crime sets my head spinning. (*OC,* 10: 174–75; *E,* 175–76)

Bataille cites Borchamps's response: "I love your head madly, and I think we shall achieve great things between us . . . rotten and corrupt as it is, I grant you" (*OC,* 10: 175; *E,* 176).

Although Bataille gives no reference, the incident with Borchamps takes place in the fifth part of *Juliette.* Oddly enough though, while the little speech by Amélie is to be found there (Sade 1966, 276), Borchamps's response, as cited, is not. In fact nowhere in Borchamps's appearances does he say anything of the sort.

Of course it is possible that Bataille found his remark in some earlier version of *Juliette,* or in some other text that mentions his name. But the standard version of *Juliette* does not include it; all Borchamps says is, in an autobiographical narrative, "Deeply moved *[ému]* myself by such a proposal, I protested to Amélie that she would have reason to be pleased with me" (Sade 1966, 276). The passage Bataille cites, wherever it may come from, is nevertheless stronger: Borchamps tells Amélie that he loves her (strictly speaking, he loves her head); in effect, he promises her that they will have a relationship based on love.

Something of the sort does take place in Sade's novel: Borchamps marries Amélie; they flee to St. Petersburg, where Borchamps commits all sorts of crimes; they live in great luxury and have what seems, strangely (for Sade), a happy conjugal life for two years. In the end no less than the Empress Catherine demands that Amélie be murdered in an orgy, and Borchamps is happy to oblige. It is clear as well that he has become bored with their relationship, since Amélie has turned out to be a bit less devoted to crime, and a bit more to faithfulness and contentment, than the libertine first suspected. Her initial speech, the one that so moved Bataille, was, he tells us, only a "raffinement de délicatesse" (refined bit of thoughtfulness) (Sade 1966, 279). It seems that Borchamps has taken Amélie at her word, whereas she had only been engaging in a bit of flattery. She finally has to be content with the end that she herself has chosen.

Important here, I think, is the fact that Bataille projects onto the figure of Amélie a kind of gracious generosity, which is returned as love by the criminal Borchamps. It is not just that Sade's characters are, as Blanchot would have it, pushing their violence to the point of total apathy toward their own selves, their own survival. Rather, an extreme devotion to crime—to, as the prewar Bataille would put it, the production of heterogeneous objects—leads, surprisingly, to a self-sacrificing generosity. The self is not simply destroyed in a whirlwind of energy; the self is destroyed through an excess of energy entailing a mortal gift of oneself in love, in crime, to the other. And that love is in some way reciprocated. The cult, the myth, of the death of God and ultimately of Man is inseparable from the gift of oneself, to the point of death.

All this is important because it indicates a direction that Bataille can follow, one that leads away from the sheer fictionality or parody of Sade (indeed, we can never be certain of parody to the extent that it is the reversible trope par excellence) toward a mythical-anthropological model in which the histories of "primitive" societies can be invoked to justify a reconceptualization of our own, modern, social order. The violence of Amélie, even though (and perhaps because) it is self-destructive, shows clearly enough that extreme "disorder" (to use an expression dear to Sade) leads not to the simple extinction of the self but also to a bond with another: the beginnings of a community. We might even say that this generosity, linked to love, is the unintended aftereffect of a violent spending of oneself to the point of self-extinction. The extreme generosity that we saw as being inseparable from the death of Man (the death of self, law, permanence) entails, unforeseen, unplanned, the establishment of a community.

The more one squanders, the more one deviates from a supreme selfishness, not just in the direction of "apathetic" self-destruction (which reaffirms the human in the very isolation of the Sadean sovereign), but in the direction of a generosity that inaugurates a social tie. This is the movement that characterizes both the death of the self (of Man) and the death of God. The will here is virtually irrelevant: it is not so much that Amélie is seeking a social bond with another through her will to a limitless death, but rather that that bond is the inevitable aftereffect of the will to the "continuity of crime," as Bataille calls it (*OC,* 10: 175; *E,* 176). That will is attained through a spending of oneself without limit, with no goal, with no desire for anything positive or constructive. Nevertheless, attaining the "continuity of crime," which is the continuity of energy that violates all arbitrary human boundaries and laws, is the attainment of a social order more profound, more basic, than that of the current bourgeois regime. It is as if the Sadean sovereign analyzed by Blanchot realizes that his or her concentration of energy and explosive death (through apathy) entails in the end not a predatory selfishness but a profound generosity—and that this generosity, through the death of God and Man, goes on to reenergize a moribund and oppressive society.

This, then, is the link between Bataille's faithful and obsessive reading and rewriting of Sade and his anthropological tendency, most clearly on display in works like "The Notion of Expenditure" and *The Accursed Share.* These works could not attempt to put forward a coherent social theory on the basis of Sade alone. What is involved, instead, is a perverse rereading of Sade in which the characters' (and by implication, Sade's) absolute selfishness morphs into a will to self-destruction and ultimately, through that, to a will to generosity. At that point one has the basis of something like a functioning society. We no longer have the apathetic selfishness of the unimaginable Sadean "Society of the Friends of Crime"—a society that would self-destruct in a matter of hours, if not minutes. Instead there is the model of a community based on giving rather than hoarding. Amélie's will to self-destruct is for that reason the first stage not toward a predatory Sadean orgy, but toward a utopian society in which people are willing to sacrifice themselves for a common myth; not one of a steady-state mode of simple conservation (of the self, of God), but a good that consists of a delirious tendency to risk the self in expenditure. Like the burst of laughter that is the end of intellectual activity, this expenditure would be the end, and not the means, of a society. It would squat in the empty space of the dead God, at the summit of (un)creation. Collective, self-risking activity would be devoted not to some higher ideal (Church, leader, State, Self) but to the activity itself—which would nevertheless have as an aftereffect a contagion (albeit ephemeral) of social cohesion and collusion. Such a society—no doubt based on a very free interpretation of Mauss's "friendly societies"[23]—would not necessarily even be violent: gift giving in principle would entail simply the sacrifice of simple utility, the squandering of the need always to reinvest what has been conserved for the purposes of future production in the name of some elusive and static principle or divinity.

Bataille's mythical utopia of generosity, if we can call it that, transgresses Sade's rigorously fictional social (de)construction. The activity, the generosity, has no goal. The self spends, it spends itself; its will to spend is an energy that risks its very existence. But the aftereffect, if we can call it that, is to open the possibility of a continuity of society: a society not of monads but of selves in contact, in "communication," selves that have broken the law and limits of selfhood, selves opened out and conjoining through wounds. The broken self, the *ipse,* like cursed matter, is inseparable from the energy that binds and that is released by, that devours, the society that Bataille envisages for the future (just as this broken self is inseparable from the scatology, the anthropology, of Bataille's heterological written project).[24] Immanent and material divinity in Bruno, the energy of godless matter in Sade, becomes in Bataille the energy of a society in which revolution means the death of an omniscient and omnipotent God (made in the image of human weakness) and the death of acquisitive, selfish, isolated Man. The aftereffect of this movement may be a social construct, a "society," that reaffirms, that opens the possibility of this expenditure, this self-sacrifice. The attainment of this community is not the result of a purposeful generous act; such a thing is inconceivable because generosity by definition is purposeless. It happens, or it doesn't, but the very fact that, according to Bataille at least, it can happen, takes us out of the realm of all-pervasive fictionality and into that of history and politics—bizarre and mythical as they might be.

We have moved, in this chapter, from questions of matter and energy to profoundly social questions of religion and ethics. But these issues are inseparable. Bataille's world, after all, is brought together in generosity and ecstasy (or dread), in and through the expenditure, the "experience" of matter and energy, affirmed "on the scale of the universe."

In the following three chapters I will sketch out how Bataille's economy, religion, and urban space are elaborated. Bataille builds on (so to speak) the set of questions derived from Bruno and Sade I have examined in this chapter: the seeming infinity of untamable energy in the universe, agitated and virulent (rather than servile) matter composing (re)creation, the self as loss of self through violent generosity, the death of God and the cult ("experience") and writing of that death; and finally, the topography of that death (the closed-off space of crime in Sade and the space of the city in Bataille). These questions are of the utmost importance not only for an understanding of Bataille but, as I argue in the second part of this book, for thinking about energy in and as the future.

2 Bataille's Ethics

Mechanized Waste and Intimate Expenditure

Given Bataille's antecedents—among them gnosticism, alchemy, Bruno, Sade—it should come as little surprise that one of his major works—*The Accursed Share* (1988a), first published in 1949, focuses on the importance of energy use and expenditure in society and nature. What may come as a surprise is that, considering the excesses of works like "The Use Value of D. A. F. de Sade," he should publish a sober historical and scientific analysis of energy and excess in traditional and modern societies. But twenty years—and World War II—separate these two works—years in which senseless destruction passed from being a desideratum of avant-garde thought to an all-too-daily event. Throughout his postwar focus on energy, Bataille nevertheless remains faithful to his earlier vision: how to see community ultimately not as the affirmation of the primacy of a narrow conservation of energy, justified through a remote, indeed inconceivable higher ideal (God, Man), but as the joyous and anguished expenditure of energy (momentarily) concentrated in beings and things. Bataille is still concerned, in other words, with the violent "transmutation" of matter at the expense of static and exhausted identities and ideals. By 1949, however, he has passed from provocation to patient analysis.

For Bataille nature and society are one and the same because both are nothing more than instances of energy concentration and waste.[1] The refocus on energy production and use has profound implications: "Man" is not so much the author of his own narrative, or the subject that experiences and acts, as "he" is the focal point of the intensification or slackening of energy flows. For this reason human life on earth must be seen as just one instance of many energy events: moments in which energy is absorbed from the sun lead to growth and reproduction but, just as important, energy is also blown off. Humans in this sense are no different from any other animals, though their wastage of energy might be more intense through its very self-consciousness. All social productions—all cultural productions—are therefore seen as modes of energy appropriation and squandering; their value or lack of value must be seen in the context of their role as conduits in the flows of energy through humans outward to the void of the universe. These flows are gifts not necessarily to other humans but to the emptiness of the sky. Gifts, or put another way, destroyed things, things whose end lies in immediate consumption not utility and deferred pleasure.

Bataille's work anticipates much recent analysis, which sees value—economic, cultural—deriving from energy inputs: humans may "produce," but their productive activity is dependent on the quantities of energy that they are capable of harnessing. Human evolution—physical and cultural—in this view is a function of the channeling of energy: taking advantage of abundant energy (derived from agricultural inputs or, later, from fossil fuels), humans reproduce and populate the earth; suffering, on the other hand, from a lack of energy, their society contracts, and they find ways to cope with eternal shortage. Surplus and shortage are thus intimately linked; each is always present in the other, and each must be recognized in its fundamental role in the preservation, extension, intensification, and ruin of the community.[2]

There are, no doubt, many ways in which the centrality of energy for life can be read. In the nineteenth century a kind of cultural pessimism was all-pervasive: since the second law of thermodynamics postulated the entropy of any given field of energy, we could then infer that any society, any life form, any planet would eventually lose the energy it had at its disposal and sink into quietude, feebleness, death. From the larger argument about energy, and the eventual fate of the sun and all other stars, commentators were quick to see a similar effect in society: the fadeout of energy led to weakness and cultural decadence. Society was on a death trip just like the sun; humans, presented in this reactionary mode, could brood over their fate but could do little to prevent it.[3]

Bataille consciously points in the opposite direction. In Bataille's view, rather than entropy, the magnificent expenditure of energy, characterized by the violence and brilliance of the sun, leads to the conclusion that energy is limitless and that the chief problem lies not in its hoarding and in the warding off of the inevitable decline, but in the glorious burn-off of the sun's surplus. In effect, the problem becomes how best to expend rather than how best to envision the consequences of shortage. For all that, Bataille is not an optimist in the conventional sense of the word because he does not link abundant energy and its glorious throughput with the placid satisfactions and order of a middle-class existence.[4]

In the 1950s there was a lot of talk about "energy too cheap to meter": the promise of the nuclear energy industry. That was good news, apparently, because it would allow us to live happy lives with a maximum number of appliances; we could always own more, always spend more, with the ultimate goal being human comfort. Growth was the name of the game—it still is—and growth in comfort was made possible when more energy was produced than needed. If energy is nothing more than the power to do work, then an unending surplus of energy meant nothing more than a continuous rise in productivity, a concomitant rise in the number of objects citizens could look forward to possessing, and the personal satisfaction associated with those objects.

Bataille too envisages a constant surplus of energy, but his energy is very different from the metered or unmetered kind. True, one can momentarily put some of Bataille's energy "to work." But there is always too much of it to be simply controlled; it always exceeds the limits of what one would be capable of devoting to some end. Bataille's energy is therefore inseparable from the wildly careening atoms of Sade or even the profoundly formless matter envisaged by Bruno. "Cursed matter," be it the charged matter studied by Durkheim, or the "base matter" of Bataille's gnosticism, or the mortal meat of Sade's "transmutation," is not only matter that is left over and so can contribute its energy to further growth; it is also matter that is burned off, which leads nowhere beyond itself, and so is dangerous, powerful, sacred.[5] Bataille's energy shoots through a charged matter that obtrudes in sacred ritual and erotic "wounds": the "share" of energy is not a resulting order but a base disorder. Such matter is in excess, not inert but virulent, threatening, turning as easily against the one who would wield the power as against a supposed victim. But along with this, the excessive, material world is "intimate," not a useful, classifiable thing, but a moment of matter that does not lead outside itself, can serve no useful purpose, is not anchored in time in such a way that it becomes a means rather than an end.

Of course no energy can be surplus in and of itself. The supposed surplus energy, too cheap to meter, of the 1950s was only surplus in relation to a power grid: there was to be so much of it that it would pulse through the power grid, illuminating backyard patios and electrically heating split-level homes for free. And the more split levels that would be built, the more available—domesticated—energy there would be to fuel the world—and so on, presumably, to infinity. Bataille's energy, however, is in surplus on another kind of grid—that of the semiotic categories of a comprehensible social system. It is what is left over when a system completes itself, when a system depends on energy in order to complete itself—but it only does so by excluding the very energy that makes its completion possible. Put another way, we can say that a social system needs to exclude a surplus of energy (hence matter) in order to constitute itself as coherent and complete. There are, in other words, limits to growth, be they external (as in an ecology) or internal (as in a social philosophy or ideology).[6] That surplus/energy, in Bataille's terminology, is "cursed," always already unusable, outside the categories of utility. It is thus not servile, not ordered or orderable. A banal example: if a rural region can produce only so much food, then its "carrying capacity" is limited; the excess human population it produces will have to be burned off in some way. A surplus of humans in a given locale will lead to contraception, warfare, celibacy, sacrifice.[7] A certain equilibrium, tentative and never truly stable, will result. Human energy, human population, will have to be lost: effort that could be spent in nonsustainable growth—producing more things that could not be absorbed—will be spent, spewed out, in other, nonproductive activities: again, war, the production of (left-hand and right-hand) sacred artifacts, "useless" art, and so on. The inevitable limit of the system—economic, ecological, intellectual—always entails a surplus that precisely *defeats* any practical appropriation. This uncontrollable and useless energy courses through the body, is the body, animating it, convulsing it: this is a threatening energy that promises death rather than any straightforward appropriation.

"Excess" matter will therefore be different in kind from its double, the "share" that can be reabsorbed into the system: the excess matter-energy will not be easily classifiable, knowable, within the parameters of the grid. It will always pose itself as a profound challenge. Against the coherent oppositions and reliable significations found operating within a given system of energy use, it constitutes a series of instances of energy in flux: never stable, never predictable, but a matrix of free energy-symbolization at the ready, to open but also to undermine the coherency of the system. Rendered docile, energy makes the system possible (society, philosophy, physics, technology); revealing itself as excessive, unconditioned, at the moment the edifice achieves its fragile summit, energy opens the abyss into which the system plunges.

Bataille's Version of Expenditure

The Accursed Share, first published in 1949, has had a colorful history on the margins of French intellectual inquiry. Largely ignored when first published, it has gone on to have an interesting and subtle influence on much contemporary thought. In the 1960s, fascination with Bataille's theory of economy tended to reconfigure it as a theory of writing: for Derrida, for example, general economy was a general *writing.* The very specific concerns Bataille shows in his work for various economic systems is largely ignored or dismissed as "muddled."[8] Other authors, such as Michel Foucault and Alphonso Lingis, writing in the wake of this version of Bataille, have nevertheless stressed, following more closely Bataille's lead, the importance of violence, expenditure, and spectacular transgression in social life.[9]

The basis for Bataille's approach can be found in the second chapter of the work "Laws of General Economy." The theory in itself is quite straightforward: living organisms always, eventually, produce more than they need for simple survival and reproduction. Up to a certain point, their excess energy is channeled into expansion: they fill all available space with versions of themselves. But inevitably, the expansion of a species comes against limits: pressure will be exerted against insurmountable barriers. At this point a species' explosive force will be limited, and excess members will die. Bataille's theory is an ecological one because he realizes that the limits are internal to a system: the expansion of a species will find its limit not only through a dearth of nourishment but also through the pressure brought to bear by other species.[10] As one moves up the food chain, each species destroys more to conserve itself. In other words, creatures higher on the food chain consume more concentrated energy. It takes more energy to produce a calorie consumed by a (carnivorous) tiger than one consumed by a (herbivorous) sheep. The ultimate consumers of energy are not so much ferocious carnivores as they are the ultimate consumers of other animals and themselves: human beings.

For Bataille, Man's primary function is to expend prodigious amounts of energy, not only through the consumption of other animals high on the food chain (including man himself) but in rituals that involve the very fundamental forces of useless expenditure: sex and death.[11] Man in that sense is in a doubly privileged position: he not only expends the most, but alone of all the animals he is able to expend *consciously.* He alone incarnates the principle by which excess energy is burned off: the universe, which is nothing other than the production of excess energy (solar brilliance), is doubled by man, who alone is aware of the sun's larger tendency and who therefore squanders consciously in order to be in accord with the overall tendency of the universe. This for Bataille is religion: not the individualistic concern with deliverance and personal salvation, but rather the collective and ritual identification with the cosmic tendency to lose.

Humans burn off not only the energy accumulated by other species but, just as important, their own energy, because humans themselves soon hit the limits to growth. Human society cannot indefinitely reproduce: soon enough what today is called the "carrying capacity" of an environment is reached.[12] Only so many babies can be born, homes built, forests harvested. Then limits are reached. Some excess can be used in the energy and population required for military expansion (the case, according to Bataille, with Islam [*OC,* 7: 83–92; *AS,* 81–91]), but soon that too screeches to a halt. A steady state can be attained by devoting large numbers of people and huge quantities of wealth and labor to useless activity: thus the large numbers of unproductive Tibetan monks, nuns, and their lavish temples (*OC,* 7: 93–108; *AS,* 93–110). Or most notably, one can waste wealth in military buildup and constant warfare: no doubt this solution kept populations stable in the past (one thinks of the endless battles between South American Indian tribes), but in the present (i.e., 1949) the huge amounts of wealth devoted to military armament, worldwide, can lead only to nuclear holocaust (*OC,* 7: 159–60; *AS,* 169–71).

This final point leads to Bataille's version of a Hegelian "absolute knowledge," one based on the certainty of a higher destruction (hence an absolute knowledge that is also a non-knowledge). The imminence of nuclear holocaust makes it clear that expenditure, improperly conceived, can threaten the continued existence of society. Unrecuperable energy, if unrecognized or conceived as somehow useful, threatens to return as simple destruction. Bataille's theory, then, is a profoundly *ethical* one: we must somehow distinguish between versions of excess that are "on the scale of the universe," whose recognition-implementation guarantee the survival of society (and human expenditure), and other versions that entail blindness to the real role of expenditure, thereby threatening man's, not to mention the planet's, survival.

This, in very rough outline, is the main thrust of Bataille's book. By viewing man as a spender rather than a conserver, Bataille manages to invert the usual order of economics: the moral imperative, so to speak, is the furthering of a "good" expenditure, which we might lose sight of if we stress an inevitably selfish model of conservation or utility. For if conservation is put first, inevitably the bottled-up forces will break loose but in unforeseen, uncontrollable, and, so to speak, untheorized ways. We should focus our attention not on an illusory conservation, maintenance, and the steady state—which can lead only to mass destruction and the ultimate wasting of the world—but instead on the modes of expenditure in which we, as human animals, should engage.[13]

But how does one go about privileging willed loss in an era in which waste seems to be the root of all evil? Over fifty years after the publication of *The Accursed Share*, we live in an era in which nuclear holocaust no longer seems the main threat. But other dangers lurk, ones just as terrifying and definitive: global warming, deforestation, the depletion of resources—and above all energy resources: oil, coal, even uranium. How can we possibly talk about valorizing heedless excess when energy waste seems to be the principal evil threatening the continued existence of the biosphere on which we depend? Wouldn't it make more sense to stress conservation, sustainability, and downsizing rather than glorious excess?

What Appears to Be Wrong with Bataille's Theory?

To think about the use value of Bataille, we must first think about the nature of energy in his presentation. For Bataille, excessive energy on the earth is natural: it is first solar (as it comes to us from the sun), then biological (as it passes from the sun to plants and animals to us), then human (as it is spent in our monuments, artifacts, and social rituals). The movement from each stage to the next involves an ever-greater disposal: the sun spends its energy without being repaid; plants take the sun's energy, convert it, and throw off the excess in their wild proliferation; and animals burn off the energy conserved by plants (carnivores are much less efficient than herbivores), all the way up the food chain. Humans squander the energy they cannot put to use in religious rituals and war. "On the surface of the globe, *for living matter in general,* energy is always in excess, the question can always be posed in terms of extravagance *[luxe],* the choice is limited to how wealth is to be squandered *[le mode de la dilapidation des richesses]*" (*OC,* 7: 31; *AS,* 23; italics Bataille's). There never is or will be a shortage of energy, it can never be used up by man or anything else, because it comes, in endless profusion, from the sun and stars.

Georges Ambrosino, a nuclear scientist and Bataille's friend, is credited in the introduction of *The Accursed Share* (*OC,* 7: 23; *AS,* 191) as the inspiration for a number of the theses worked out in the book. In some unpublished "notes preliminary to the writing of *The Accursed Share*" (*OC,* 7: 465–69), Ambrosino sets out very clearly some of the ideas underlying Bataille's work:

> We affirm that *the appropriated energies produced during a period are superior in quantity to the appropriated energies that are strictly necessary to their production.*
>
> For the rigor of the thesis, it would be necessary to compare the appropriated energies of the same quality. The system produces all the appropriated energies that are necessary to it, it produces them in greater quantities than are needed, and finally it even produces appropriated energies that its maintenance at the given level does not require.
>
> In an elliptical form, but more striking, we can say that *the energy produced is superior to the energy necessary for its production.* (*OC,* 7: 469)[14]

Most striking here is the rather naive faith that, indeed, there always will be an abundance of refinable, usable energy and that spending energy to get energy inevitably results in an enormous surplus of energy—so much that there will always be a surplus, "greater quantities than are needed." Ambrosino, in other words, projects a perpetual surplus of energy return on energy investment (EROEI).[15] One can perhaps imagine that a nuclear scientist, in the early days of speculation about peaceful applications of atomic energy, might have put it this way. Or a petroleum geologist might have thought the same way, reflecting on the productivity of the earth shortly after the discovery of a giant oil field.[16] Over fifty years later it is much harder to think along these lines.

Indeed, these assumptions are among those most contested by current energy theorists and experts. First, we might question the supposition that since all energy in the biosphere ultimately derives from the sun, and the sun is an inexhaustible source of energy (at least in relation to the limited life spans of organisms), there will always be a surplus of energy *for our use.* The correctness of this thesis depends on the perspective from which we view the sun's energy. From the perspective of an ecosystem—say, a forest—the thesis is true: there will always be more than enough solar energy so that plants can grow luxuriantly (provided growing conditions are right: soil, rainfall, etc.) and in that way supply an abundance of biomass, the excess of which will support a plethora of animals and, ultimately, humans. All living creatures will in this way always absorb more energy than is necessary for their strict survival and reproduction; the excess energy they (re)produce will inevitably, somehow, have to be burned off. There will always be *too much* life.

If we shift perspective slightly, however, we will see that an excess of the sun's energy is not always available. It is (and will continue to be) extremely difficult to achieve a positive energy return directly from solar energy.[17] As an energy form, solar energy has proven to be accessible primarily through organic (and fossilized) concentration: wood, coal, and oil. In human society, at least as it has developed over the last few millennia, these energy sources have been tapped and have allowed the development of human culture and the proliferation of human population. It has often been argued that this development/proliferation is not due solely to technological developments and the input of human labor; instead, it is the ability to utilize highly concentrated energy sources that has made society's progress possible. Especially in the last two hundred years, human population has expanded mightily, as has the production of human wealth. This has been made possible by the energy contributed to the production and consumption processes by the combustion of certain fuels in ever more sophisticated mechanical devices: first wood and then coal in steam engines, and then oil and its derivatives (including hydrogen, via natural gas) in internal combustion engines or fuel cells. Wealth as it has come to be known in the last three hundred or so years, in other words, has its origins not just in the productivity of human labor and its ever more sophisticated technological refinements, as both the bourgeois and Marxist traditions would argue, but in the energy released from (primarily) fossil fuels through the use of innovative devices. In the progress from wood to coal and from coal to oil, there is a constant progression in the amount of quantifiable and storable energy produced from a certain mass of material.[18] Always more energy, not necessarily efficiently used; always more goods produced, consumers to consume them, and energy-based fertilizers to produce the food needed to feed them. The rise of civilization as we know it is tied directly to and is inseparable from the type of fuels used to power and feed it—and the quantities of energy derived from those fuels at various stages of technology.[19]

Certainly Bataille, following Ambrosino, would see in this ever-increasing energy use a continuation—but on a much grander scale—of the tendency of animals to expend energy conserved in plant matter. Indeed, burning wood is nothing more than that. But the fact remains that by tapping into the concentrated energy of fossil fuels, humans have at their disposal (ancient) solar energy—derived from fossil plants (coal) and algae (oil)—in such a concentrated form that equivalent amounts of energy could never be derived from solar energy alone.[20]

In a limited sense, then, Bataille and Ambrosino are right: all the energy we use ultimately derives from the sun. There is always more of it than we can use. Where they seem to be wrong is that they ignore the fact that for society as we know it to function, with our attendant leisure made possible by "energy slaves," energy derived from fossil fuels, with their high EROEI, will be necessary for the indefinite future.[21] There is simply no other equally rich source of energy available to us; moreover, no other source will likely be available to us in the future.[22] Bataille's theory, on the other hand, ultimately rests on the assumption that energy is completely renewable, there will always be a high EROEI, and we need not worry about our dependence on finite (depletable) energy sources. *The Accursed Share* for this reason presents us with a strange amalgam of awareness of the central role energy plays in relation to economics (not to mention life in general) and a willful ignorance concerning the social-technological modes of energy delivery and use, which are far more than mere technical details. We might posit that the origin of this oversight in Bataille's thought is to be found in the economic theory, and ultimately philosophy, both bourgeois and Marxist, of the modern period, where energy resources and raw materials for the most part do not enter into economic (or philosophical) calculations, since they are taken for granted: the earth makes human activity possible, and in a sense we give the earth meaning, dignity, by using resources that otherwise would remain inert, unknown, insignificant (one thinks of Sartre's "in-itself" here). Value has its origin, in this view, not in the "natural" raw materials or energy used to produce things but in human labor itself. Bataille merely revises this model by characterizing human activity, in other words production, as primarily involving gift-giving and wasting rather than production and accumulation.

We can argue, then, that solar energy is indeed always produced, always in excess (at least in relation to the limited life spans of individuals and even species); but it is fossil fuels that best conserve this energy and deliver it in a rich form that we humans can most effectively use. Human progress has been so explosive that these energy inputs have come to seem infinite and then have become invisible. Unfortunately, fossil fuels can be depleted, indeed are in the process of being depleted.

Why is this important in the context of Bataille? For a very simple reason: if Bataille does not worry about energy cost and depletion, he does not need to worry about energy conservation. Virtually every contemporary commentator on energy use sees only one short-term solution: conservation. Since fossil fuels are not easily replaceable by renewable sources of energy, our only option is to institute radical plans for energy conservation—or risk the complete collapse of our civilization when, in the near future, oil, coal, and natural gas production decline and the price of fuel necessarily skyrockets.[23] Some commentators, foreseeing the eventual complete depletion of fossil energy stores, predict a return to feudalism (Perelman 1981) or simply a quasi-Neolithic state of human culture, with a radically reduced global population (Price 1995).

Without a theory of depletion, Bataille can afford to ignore conservation in all senses: not only of resources and energy but also in labor, wealth, and so on. He can also ignore (perhaps alarmist) models of cultural decline. In Bataille's view, there will always be a surplus of energy; the core problem of our civilization is how we use up this excess. We need never question the "energy slaves" inseparable from our seemingly endless waste. Nor will there need to be any consideration of the fact that these energy slaves may very well, in the not-so-distant future, have to be replaced by real, human slaves.

Bataille, Depletion, and Carrying Capacity

Steven A. LeBlanc's book *Constant Battles: The Myth of the Peaceful, Noble Savage* (2003) would seem, at least at first, to pose an insuperable challenge to Bataille's view of wealth, expenditure, excess, and the social mechanisms that turn around them. LeBlanc's larger argument is that warfare in all societies—hunters and gatherers, farmers, as well as industrialized, "modern" societies—arises from competition for increasingly scarce resources as the carrying capacity of the land decreases. It should be stressed that carrying capacity[24] is linked to population growth: the latter is never stable, and up to a point the land can support an increasing number of individuals. There is, however, an inverse relation between population and carrying capacity: the limits of the latter are rapidly reached through a burgeoning population, and a higher population depletes the productive capacity of the environment, thereby making the revised carrying capacity inadequate even for a smaller population. But as carrying capacity is threatened, many societies choose warfare, or human sacrifice, rather than extinction (LeBlanc 2003, 177–78, 195).

I stress the importance of LeBlanc's thesis—that violent conflict arising out of ever-growing population pressures and diminishing carrying capacity of the environment characterizes all developmental levels of human society—because it highlights another apparent weakness of Bataille's theory. LeBlanc would argue that there is no model of what we held so dear in the 1960s: a noble savage—Native American, Tibetan, or whoever—who is or was "in harmony with the environment." Bataille's theory would seem to posit just such a harmony, albeit one that involves the violence of sacrifice rather than the contentment of the lotus-eater. Man in his primitive state was in harmony not with the supposed peace of Eden but with the violence of the universe, with the solar force of blinding energy:

> The naïve man was not a stranger in the universe. Even with the dread it confronted him with, he saw its spectacle as a festival to which he had been invited. He perceived its glory, and believed himself to be responsible for his own glory as well. (*OC,* 7: 192)

While LeBlanc's theory of sacrifice is functional—he is concerned mainly with how people use sacrifice in conjunction with warfare to maximize their own or their group's success—Bataille's theory is religious in that he is concerned with the ways in which people commune with a larger, unlimited, transcendent reality. But in order to do so they must apparently enjoy an unlimited carrying capacity.

And yet, if we think a bit more deeply about these two approaches to human expenditure (both LeBlanc and Bataille are, ultimately, theorists of human violence), we start to see notable points in common. Despite appearing to be a theorist of human and ecological scarcity, LeBlanc nevertheless presupposes one basic fact: there is always a tendency for there to be *too many* humans in a given population. Certainly populations grow at different rates for different reasons, but they always seem to outstrip their environments: there is, in essence, always an excess of humans that has to be burned off. Conversely, Bataille is a thinker of limits to growth, precisely because he always presupposes a limit: if there were no limit, after all, there could be no excess of anything (yet the limit would be meaningless if there were not always already an excess: the excess opens the possibility of the limit). As we know, for Bataille too there is never a steady state: energy (wealth) can be reinvested, which results in growth; when growth is no longer possible, when the limits to growth have been reached, the excess must be destroyed. If it is not, it will only return to cause us to destroy ourselves: war.

> For if we aren't strong enough to destroy, on our own, excessive energy, it cannot be used; and, like a healthy animal that cannot be trained, it will come back to destroy us, and we will be the ones who pay the costs of the inevitable explosion. (*OC,* 7: 31; *AS,* 24)

In fact, Bataille sounds a lot like LeBlanc when he notes, in *The Accursed Share,* that the peoples of the "barbarian plateaus" of central Asia, mired in poverty and technologically inferior, could no longer move outward and conquer other adjacent, richer areas. They were, in effect, trapped; their only solution was the one that LeBlanc notes in similar cases: radical infertility. This, as noted by Bataille, was the solution of the Tibetans, who supported an enormous population of infertile and unproductive monks (*OC,* 7: 106; *AS,* 108).

Bataille does, then, implicitly face the question of carrying capacity. Perhaps the ultimate example of this is nuclear war. The modern economy, according to Bataille, does not recognize the possibility of excess and therefore limits; the Protestant, and then the Marxist, ideal is to reinvest all excess back into the productive process, always augmenting output in this way. "Utility" in this model ends up being perfectly impractical: only so much output can be reabsorbed into the ever-more-efficient productive process. As in the case with Tibet, ultimately the excess will have to be burned off. This can happen either peacefully, through various postcapitalist mechanisms that Bataille recommends, such as the Marshall Plan, which will shift growth to other parts of the world, or violently and apocalyptically through the ultimate in war: nuclear holocaust. One can see that, in the end, the world itself will be *en vase clos,* fully developed, with no place for the excess to go. The bad alternative—nuclear holocaust—will result in the ultimate reduction in carrying capacity: a burned-out, depopulated earth.

> Humanity is, at the same time, through industry, which uses energy for the development of the forces of production, both a multiple opening of the possibilities of growth, and the infinite faculty for burnoff in pure loss *[facilité infinie de consumation en pure perte].* (*OC,* 7: 170; *AS,* 181)

Modern war is first of all a renunciation: one produces and amasses wealth in order to overcome a foe. War is an adjunct to economic expansion; it is a practical use of excessive forces. And this perhaps is the ultimate danger of the present-day (1949) buildup of nuclear arms: armament, seemingly a practical way of defending one's own country or spreading one's own values, in other words, of growing, ultimately leads to the risk of a "pure destruction" of excess—and even of carrying capacity. In the case of warfare, destructiveness is masked, made unrecognizable, by the appearance of an ultimate utility: in this case the spread of the American economy and the American way of life around the globe. Paradoxically, there is a kind of self-consciousness concerning excess, in the "naïve" society—which recognizes expenditure for what it is (in the form of unproductive glory in primitive warfare)—and a thorough ignorance of it in the modern one, which would always attempt to put waste to work ("useful" armaments) even at the cost of wholesale destruction.

Bataille, then, like Le Blanc, can be characterized as a thinker of society who situates his theory in the context of ecological limits. From Bataille's perspective, however, there is always too much rather than too little, given the existence of ecological ("natural") and social ("cultural") limits. The "end" of humankind, its ultimate goal, is thus the destruction of this surplus. While Le Blanc stresses war and sacrifice as a means of obtaining or maintaining what is essential to bare human (personal, social) survival, Bataille emphasizes the maintenance of limits and survival as mere preconditions for engaging in the glorious destruction of excess. The meaning of the limit and its affirmation is inseparable from the senselessness of its transgression in expenditure *(la dépense).* By seeing warfare as a mere (group) survival mechanism, Le Blanc makes the same mistake as that made by the supporters of a nuclear buildup; he, like they, sees warfare as practical, serving a purpose, and not as the sheer burn-off it really is.

If, however, our most fundamental gesture is the destruction of a surplus, the production of that surplus must be seen as subsidiary. Once we recognize that everything cannot be saved and reinvested, the ultimate end (and most crucial problem) of our existence becomes the disposal of excess wealth (concentrated, nonusable energy). All other activity leads to something else, is a means to some other end; the only end that leads nowhere is the act of destruction by which we may—or may not—assure our (personal) survival (there is nothing to guarantee that radical destruction—*consumation*—does not turn on its author). We work in order to spend. We strive to produce sacred (charged) things, not practical things. Survival and reproduction alone are not the ultimate ends of human existence. We could characterize Bataille for this reason as a thinker of ecology who nevertheless emphasizes the primacy of an ecstatic social act (destruction). By characterizing survival as a means not an end (the most fundamental idea in "general economy"), expenditure for Bataille becomes a limitless, insubordinate act—a real *end* (that which does not lead outside itself). I follow Bataille in this primacy of the delirium of expenditure over the simple exigency of personal or even social survival (Le Blanc). This does not preclude, however, a kind of ethical aftereffect of Bataille's expenditure: *survival for this reason can be read as the fundamentally unintentional consequence of expenditure rather than its purpose.* Seeing a nuclear buildup as the wrong kind of expenditure—because it is seen as a means not an end—can lead, in Bataille's view, to a rethinking of the role of expenditure in the modern world and hence, perhaps, the world's (but not modernity's) survival.

Limits

Carrying capacity poses a limit to growth: a society can destroy the excess through sacrifice, infanticide, ritual, festival; or excess can be put to work through the waging of war, in which case carrying capacity may be expanded through the appropriation of another society's land. War too, however, shows some elements of religious, ruinous expenditure in that it entails, as does sacrifice, glory. Especially in modern times, war also brings with it the possibility of defeat: in that case there is no glory, and certainly no possibility of the expansion of carrying capacity. Indeed, as in the case of nuclear holocaust, societies run the risk of completely obliterating—wasting—the carrying capacity of their land.

In accord with Bataille's implicit ethical model, one can argue that the limits imposed by carrying capacity evoke two possible responses from societies. First, a society can recognize limits. Here, paradoxically, one violates limits, consciously transgresses them, so to speak, by recognizing them. Through various forms of ritual expenditure one ultimately respects limits by symbolically defying the very principle of conservation and measured growth—of, in other words, limits. "Spending without reserve" is the spending of that which cannot be reinvested because of the limit, and yet the very act of destruction is the transgression of the logic of the limit, which would require, in its recognition, a sage and conservative attentiveness to the dangers of excessive spending. If there is a limit to the production of goods and resources, however, we best respect and recognize that limit through its transgression—through, in other words, the destruction of precious but unusable energy resources. To attempt to reinvest, or put to use, the totality of those resources, to guarantee maximum productivity and growth, would only ignore the limit (rather than transgressing it), thereby eventually lowering the limit if not eliminating it entirely (elimination of carrying capacity, ecological destruction, desertification).[25] For this reason, a theory of expenditure is inseparable from, is even indistinguishable from, a theory of depletion.

Such an affirmation—of limits and expenditure—entails a *general* view of economy and, we might add, ecology. In positing such a respect for limits through their transgression, we forgo an individual concern, which would customarily be seen as the human one (but which is not, in Bataille's view): a concern with personal survival, enrichment, and advancement. From a larger perspective, we forego the needs of Man as a species or moral category (or the needs of God as Man's moral proxy). The supremacy of self-interest is tied for Bataille to the simple ignorance of limits: not their transgression, but their heedless violation. In the case of transgression of limits, we risk what might be personally comfortable or advantageous in order to attain a larger "glory" that is tied to unproductive expenditure and entails a possible dissolution of the self. From a general perspective, this expenditure is (as Bataille would say) on the scale of the universe; it must also be, in principle, on the scale of the carrying capacity of a given landscape or ecology (else the expenditure would very quickly cancel itself out).

This version of limits and their transgression can be associated with Bataille's conception of eroticism. What separates humans from the animals, according to Bataille, is the interdiction of "immediate, unreserved, animal pleasure *[jouissance]*" (*OC,* 8: 47). Decency, the rules against sexual expression, incest, and intense pleasure that characterize human society are fundamental to an organized society. But the human is not exclusively to be found in the interdiction: its ultimate "self-consciousness" is derived through the ecstatic transgression of that interdiction. Interdiction is an aftereffect of transgression, just as conservation is an aftereffect of expenditure (we produce and conserve in order to expend). What ultimately counts for us as humans (for us to be human) is an awareness of the necessity of expenditure (including that of our own death)—an awareness that animals lack.

> Of course, respect is only the detour of violence. On the one hand, respect orders the humanized world, where violence is forbidden; on the other, respect opens to violence the possibility of a breakout into the domain where it is inadmissible. The interdiction does not change the violence of sexual activity, but, by founding the *human* milieu, it makes possible what animality ignored: the transgression of the rule. . . .
>
> What matters is essentially that a milieu exists, no matter how limited, in which the erotic aspect is unthinkable, and moments of transgression in which eroticism attains the value of the greatest overthrow *[renversement].* (*OC,* 8: 47–48; italics Bataille's)

Eroticism, the general or collective experience of transgression, is impossible without the knowledge of human limits, interdictions. In the same way, we can say that the destruction of excess in an economy is only "on the scale of the universe" if it maintains and respects limits. We could even go beyond this and say that the maintenance of those limits, the carrying capacity in today's terminology, is only possible through the ritual, emotionally charged destruction of excess wealth (and not its indefinite, seemingly useful, but indifferent reproduction), just as interdictions are only meaningful, and therefore maintainable, when they are periodically transgressed.

The only other approach to limits, as I have indicated, is to ignore them: the consumption of scarce resources should go on forever; growth is limitless. In the realm of eroticism, this would be either to be entirely unaware of moral limits (interdictions)—as are animals—or on the other hand, to see limits as so absolute that no meaningful transgression can take place; in this case all eroticism would be so minor, so secondary, that no intimate relation between interdiction and eroticism could be imagined, and no dependence of interdiction on the transgressive expression of eroticism could be conceived. In this case limits would be so overwhelming that they would not even be limits: in effect one could not violate them since they would be omnipresent, omnipotent. Their transgression would be inconceivable (to try to violate them would simply manifest one's own degeneracy or evil, one's status outside the community, in an asylum or hell). Not coincidentally, this position is that of a religious-social orientation in which flamboyant expenditure—sexual, religious, phantasmic—is inconceivable, or unworthy of conception, and in which all excess must therefore be reinvested in material productive processes (even eroticism is subordinated to the production of more people): Calvinism, the Protestant ethic, various fundamentalisms, and so on. This is the narrow view, that of the restricted economy, the economy of the "individual":

> Each investor demands interest from his capital: that presupposes an unlimited development of investment, in other words the unlimited growth of the forces of production. Blindly denied in the principle of these essentially productive operations is the not unlimited but considerable sum of products consumed in pure loss *[en pure perte]*. (*OC,* 7: 170; *AS,* 182)

This restricted economy, which hypostatizes limits (moral, personal) only ultimately to ignore them or degrade them, is the economy that values war as a mode of expansion (typified, for Bataille at least, by Islam) and as utility (self-defense, deterrence, mutually assured destruction). The limit is ignored in the restricted economy only at the risk of reimposition of an absolute limit, cataclysmic destruction, or ecological collapse (nuclear holocaust, the simple elimination of carrying capacity).

Bataille's ethics, then, entail a choice between these two alternatives: recognition of limits through the affirmation of expenditure in a general economy, and the ignorance of limits through a denial of expenditure in a closed or restricted economy. The first entails the affirmation of glorious pleasure, sacred matter and energy, and anguish before death, while the latter entails the ego-driven affirmation of utility and unlimited growth with all the attendant dangers (the untheorized and quite sudden imposition of the limits to growth).

The irony in all this is that the first, transgressive, and "human" ethics will inevitably be sensitive to ecological questions—respectful of carrying capacity—through its very affirmation of loss. The second, attempting to limit severely or do away with waste and thereby affirm the particular interests of an individual, a closed social group, or a species (Man) in the name of "growth," will only universalize the wasting—the ultimate destruction—of the carrying capacity that serves as the basis of life. Conservation is therefore a logical aftereffect of expenditure; we conserve in order to expend. In other words, we conserve not to perpetuate our small, monadic existences or the putative centrality of our species, but rather to make possible a larger generosity, a larger general economy that entails the transgression (in *angoisse*) of our narrow, selfish "practicality," our limitedness (i.e., the inevitable postponement of pleasure).

By expending we conserve. Bataille's utopian ethics foresees a society that creates, builds, and grows in and through loss. Bataille thereby affirms the continuation of a human collectivity whose humanity is inseparable from that general—collective and ecstatic—expenditure. Inseparable, in other words, from a loss of the very selfish fixation on knowledge, authority, and even comforting immortality with which the word humanity is usually associated. The raison d'être of society, so to speak, will lie in the very unreasoned logic of its excessive and transgressive expenditure. This highest value will be maintained and known through recognition of limits, which is ultimately reasonable but to which the act of expenditure nevertheless cannot be reduced (because the affirmation of limits entails their transgression at the "highest" point of development and knowledge).

The Duality of Expenditure and the Object

Bataille's model in *The Accursed Share* ultimately depends on a distinction between types of expenditure and what we might call the modes of being associated with each type. This is significant because much of Bataille's analysis entails a critique of the confusion between different types of expenditure and economy: the "restrained" and the "general." Indeed, Bataille would argue that many of our current ills under capitalism derive from the confusion between the two realms; a Bataillean ethics would work to separate them.

First, "good" expenditure. Bataille associates it with an uncontrollable "élan": "riches prolong the burst of the sun and invoke passion"; "it's the return of the breadth of living to the truth of exuberance" (*OC,* 7: 78; *AS,* 76). Here again we have the passions unleashed by a naive intimacy with the sun and the profound workings of the universe. But this intimacy is inseparable from the violence of enthusiasm.

Contrary to the world of work, the world of expenditure entails spending without regard for the future, affirmation of ecstasy now, and the refusal of things *(les choses)* that only serve a purpose and that contribute only to one's own personal security and satisfaction (profit). Thus Bataille's theory is not only an economic one but an ethical one that criticizes the affirmation of *self.*

As we have seen, however, this affirmation does not serve to deny what is usually, and perhaps wrongly, associated with the self: pleasure. For this reason Bataille proposes a subject, which, in its habitation of an intimate world, refuses the stable and reasonable order of things in order to enter into a profound communication with others and with the universe. This communication, this intimacy, this generosity, entails a kind of relation that is radically different from the use of a seemingly stable thing to achieve a purpose. In *The Accursed Share,* Bataille writes:

> The *intimate* world is opposed to the *real* as the measureless is to measure, as madness is to reason, as drunkenness is to lucidity. There is only measure in the object, reason in the identity of the object with itself, lucidity in the direct knowledge of objects. The world of the subject is night: this moving, infinitely suspect night, which, in the sleep of reason, *engenders monsters. I propose, concerning the free subject, which is not at all subordinate to the "real" order and which is occupied only in the present, that in principle madness itself can give us only an adulterated idea.* (*OC,* 7: 63; *AS,* 58; italics Bataille's)

In spite of this emphasis on the subject, it should be stressed that Bataille is attempting to put forward a concept of the instant and of experience—if those words have any meaning at all—which exit from the personal, individual realm; the very notion of a general economy means that individual, isolated interest is in principle left behind, and instead a larger perspective is embraced, one in which the individual's concerns and worries are no longer paramount. Replacing them are the larger energy flows of the death-bound, erotic subject, of society in the grip of collective frenzy or revolt, and of the universe in the unrecoverable energy of a myriad of stars.

Having said all this, one should stress that this Bataillean ideal—for that's what it is, really—is itself already double, mixed with a recognition of the other reality. The *angoisse*—anguish, dread—before this "inner experience" is a human cut of sense, meaning, and purpose with which one engages when one comes to "face death." "Joy before death" is not separable from a dread that serves to instill a human meaning in an otherwise cosmic, but limitless and hence nonhuman, event. Without dread, in other words, the "subject" merely melds with the ambient surroundings, like an animal. It is dread—which includes the very human knowledge of the limit, of death—that serves to demarcate the event and thus give it meaning. A limit that is recognized, affirmed, at the instant of its transgression.

Meaning? Does that mean it is "significant"? For what? For some useful purpose? Not entirely. Dread entails a recognition of limits, of course, but also their defiant overcoming; much like Mozart and Da Ponte's unstoppable Don Giovanni, the "subject" recognizes and affirms the limit only to overcome it, in defiance. In the same way, transgression inevitably entails an affirmation, along with an overcoming, of interdiction. Sacrifice entails dread: it is "communication"—but communication *of* dread (*OC*, 7: 518).

Bataille also makes it clear that dread is intimately tied to sense, even to reason. As he puts it in some unpublished notes to *The Limits of the Useful* (written shortly before *The Accursed Share*):

> To anyone who wants glory, the inevitable dread must first be shown. Dread distances only impertinence *[outrecuidance]*. The danger of "strong feelings" is that one will speak of them before experiencing them: one tries to provoke them by verbal violence, but one only ends up introducing violence without force. (*OC*, 7: 512)

Bataille goes on to speak of the ancient Mexicans, but their "reality" only underscores the need for an "anguished *[angoissé]* and down to earth *[terre à terre]* research." A "slow rigor" is required to "change our notion of ourselves and of the Universe" (*OC*, 7: 512).

All this is ultimately important because it shows us the dual nature of Bataille's project. It is not just an affirmation of death, madness, wild destruction, and the leap into the void. These terms, associable with excess, expenditure, indicate "events" or "experiences" (for want of better words) moved toward—they can never simply be grasped, attained—what would seem to be their contrary: interdiction, the limit, down-to-earth research.

Transgression would not be transgression without the human limit of meaning—of interdiction, of scarcity—against which it incessantly moves. Bataille's method is not that of the raving madman but of the patient economist, writing against a "closed" economy, and of the Hegelian, writing against a narrow consciousness that would close off ecstasy, expenditure, and loss. Indeed, the final point Bataille wishes to reach is a higher "self-consciousness," not of a stable and smug universal awareness but of a knowledge facing, and impossibly grasping, a general economy of loss—in dread. Thus Bataille can write of a self-consciousness that "humanity will finally achieve in the lucid vision of a linkage of its historical forms" (*OC,* 7: 47; *AS,* 41).

A very particular self-consciousness, then, linked to a very peculiar concept of history. A self-consciousness, through a "slow rigor," that grasps "humanity" not as a stable or even dynamic presence, but as a principle of loss and destruction. A history not of peak moments of empire, democracy, or class struggle, but as exemplary instances of expenditure. And a future not in absolute knowing, but in a finally utopian "non-knowledge," "following the mystics of all periods," as Bataille puts it in the final footnote to *The Accursed Share* (*OC,* 7: 179; *AS,* 197). But he then goes on to add, about himself: "but he is no less foreign to all the presuppositions of various mysticisms, to which he opposes only the lucidity of *self-consciousness*" (italics Bataille's).

So there is, then, what we might call a good duality in Bataille. In fact, the "accursed share" is itself, for want of a better term, doubled: it entails and presupposes limits, dread, self-consciousness, language (*OC,* 7: 596–98), along with madness, "pure loss," death. The accursed share, in other words, entails the duality of transgression, in anguish *(l'angoisse),* of the recognized and ultimately affirmed limits of self, body, and world.

But the same thing could be said, again for want of a better term, of the various ways this "part" is diluted or betrayed: what we might call, to differentiate it, "bad duality" (in contradistinction to the "good" duality of the transgression, in *angoisse,* of the recognized limits of self, body, and world).

"Bad duality," as I crudely put it, is the indulgence in expenditure out of personal motives: to gain something for oneself (glory, social status) or for one's social group or nation (booty, territory, security). From the chief who engages in potlatch, all the way to the modern military planners of nuclear war—all conceive of a brilliant, radical destruction of things as a *useful* contribution: to one's own social standing, to the position or long-term survival of one's own society.

And yet, for all that, Bataille recognizes a kind of devolution in warfare: earlier (sacrificial) war and destructive gift-giving still placed the emphasis on a spectacular and spectacularly useless destruction carried out on a human scale. Later warfare, culminating in nuclear war, heightens the intensity of destructiveness while at the same time reducing it to the status of simple implement: one carries out destructive acts (e.g., Hiroshima) to carry out certain useful policy goals. "Primitive" war, then, was closer to what I have dubbed "good" duality.

Implicit in Bataille's discussion of war, from the Aztecs to the Americans, is the loss of intimacy. Aztec war was thoroughly subordinated, both on the part of victor and vanquished, to the exigencies of passion; as time went on, it seems that martial glory came to be associated more and more with mere rank. Self-interest replaced the "intimate," exciting destruction of goods and life. Modern nuclear war is completely devoid of any element of transgression or dread; it is simply mechanized murder, linked to some vague political or economic conception of necessity. Ultimately, for this reason, war in Bataille's view must be replaced by a modern version of potlatch in which one nation-state (the United States) gives without counting to others (the Europeans, primarily).

Modern war remains, for all that, an example of mankind's tendency to expend. It is merely an extreme example of an inability to recognize *dépense* for what it is. It thereby constitutes a massive failure of self-consciousness: "bad duality" as the melding of the "tendency to expend" with the demand for utility and self-interest.

Something, however, is missing in Bataille's analysis. This steady progression in types of warfare, while signaling the difference between what we might call "intimate" war (the Aztecs) and utilitarian war (the World Wars), nevertheless does tend to conflate them, in a very specific way. They are all seen as moments in which humanity plays the role of the most efficient destroyer, the being at the top of the food chain that consumes—in both senses of the word—the greatest concentrations and the greatest quantities of energy. Ultimately the difference between Aztec war and American war is exclusively one of self-consciousness; ironically, it was the Aztecs who, in their sacrificial/militaristic orgies, were in closer touch with and had greater awareness of the nature of war. The Americans, quantitatively, might be the greater consumers, but their knowledge of what they are doing is minimal (only the Marshall Plan, augmented through a reading of Bataille, would solve that problem).[26]

What is not discussed is the nature of the destruction itself. Bataille never considers that contemporary *dépense* is not only greater in *quantity* but is different in *quality*. How is it that mankind has gone from the relatively mild forms of destruction practiced by the Aztecs—mountains of skulls, to be sure, but still, relatively speaking, fairly harmless—to the prospect of the total devastation of the earth? Why has destruction been amplified to such a degree? Does it change the very nature of the expenditure carried out by modern societies?

The answer, I think, is to be found in the nature of the consumption itself. Bataille in effect makes the same mistake that traditional economists make concerning the origin of value: that it is to be found primarily in human labor. If, however, we see the skyrocketing of the creation of value in the last two centuries to be attributable not solely to inputs of human labor (muscle and brain power) but above all to the energy derived from fossil fuels (as Beaudreau [1999] claims), we will come to understand that the massive increase in mankind's capacity to waste is attributable not only to, say, technical innovation, the more efficient application of human labor, genius, and so on, but to the very energy source itself. The Aztecs, like many other traditional societies, derived their energy from muscle power: that of animals, slaves, and, in warfare, nobles. Destruction, like production, entailed an expenditure of energy derived from very modest sources: calories derived from food (solar energy), transformed by muscle, and applied to a task. We might call this energy (to modify a Bataillean usage) and its destruction *intimate:* that is, its production and expenditure are on a human scale, and are directly tied to a close bodily relation with things. This relation implies a corporeal engagement with and through an energy that *cannot* be put to use, that fundamentally defies all appropriation. Just as intimacy for Bataille implies a passionate involvement with the thing—primarily its *consumation,* its burn-off, the intense relation with a thing that is not a thing (as opposed to *consommation,* in the sense of everyday purchase, use, and wastage)—so in this case, having to do with the production and destruction of value, my muscle power assures that my relation to what I make or destroy will be passionate. A hand tool's use will entail physical effort, pain, pleasure, satisfaction, or anguish. It will be up close and personal. The same will go for the destruction of the utility of that tool; there will be a profound connection between "me" and the destruction of the thing-ness of the tool.[27] By extension, the utility, "permanence," and thus the servility of my self will be put in question through an intimate connection ("communication") with the universe via the destroyed or perverted object or tool.

Just as there are two energetic sources of economic value, then—muscle power and inanimate fuel power—so too there are two kinds of expenditure. The stored and available energy derived from fossil or inanimate fuel expenditure, for production or destruction, is different in quality, not merely in quantity, from muscular energy. The latter is profoundly more and other than the mere "power to do work." No intimacy (in the Bataillean sense) can be envisaged through the mechanized expenditure of fossil fuels. The very use of fossil and nonorganic fuels—coal, oil, nuclear—implies the effort to maximize production through quantification, the augmentation of the sheer quantity of things. Raw material becomes, as Heidegger put it, a standing reserve, a measurable mass whose sole function is to be processed, used, and ultimately discarded.[28] It is useful, nothing more (or less), at least for the moment before it is discarded; it is related to the self only as a way of aggrandizing the latter's stability and position. There is no internal limit, no *angoisse* or pain before which we shudder; we deplete the earth's energy reserves as blandly and indifferently as the French revolutionaries (according to Hegel) chopped off heads: as if one were cutting off a head of cabbage. "Good" duality has completely given way to "bad." As energy sources become more efficiently usable—oil produces a lot more energy than does coal, in relation to the amount of energy needed to extract it, transport it, and dispose of waste (ash and slag)—more material can be treated, more people and things produced, handled, and dumped. Consequently more food can be produced, more humans will be born to eat it, and so on (the carrying capacity of the earth temporarily rises). And yet, under this inanimate fuels regime, the very nature of production and above all destruction changes. Even when things today are expended, they are wasted under the sign of efficiency, utility. This very abstract quantification is inseparable from the demand of an efficiency that bolsters the position of a closed and demanding subjectivity. We "need" cars and SUVs, we "need" to use up gas, waste landscapes, forests, and so on: it is all done in the name of the personal lifestyle we cannot live without, which is clearly the best ever developed in human history, the one everyone necessarily wants, the one we will fight for and use our products (weapons) to protect. We no longer destroy objects, render them intimate, in a very personal, confrontational potlatch; we simply leave items out for the trash haulers to pick up or have them hauled to the junkyard. Consumption *(la consommation)* in the era of the standing reserve, the framework *(Ge-Stell),* entails, in and through the stockpiling of energy, the stockpiling of the human: the self itself becomes an element of the standing reserve, a thing among other things. There can hardly be any intimacy in the contemporary cycle of production-consumption-destruction, the modern and degraded version of expenditure. As Bataille put it, concerning intimacy:

> Intimacy is expressed only under one condition by the *thing [la chose]:* that this *thing* fundamentally be the opposite of a thing, the opposite of a product, of merchandise: a burn-off *[consumation]* and a sacrifice. Since intimate feeling is a burn-off, it is burning-off that expresses it, not the *thing,* which is its negation. (*OC,* 7: 126; *AS* 132: italics Bataille's)

War, too, reflects this nonintimacy of the thing: fossil fuel and nuclear-powered explosives and delivery systems make possible the impersonal destruction of lives in great numbers and at a great distance. Human beings are now simply quantities of material to be processed and destroyed in wars (whose purpose is to assure the continued availability of fossil fuel resources). Killing in modern warfare is different in kind from that carried out by the Aztecs. All the sacrificial elements, the elements by which the person has been transformed in and through death, have disappeared.

Bataille, then, should have distinguished more clearly between intimate and impersonal varieties of useless squandries when it came to his discussion of the Marshall Plan.[29] (In the same way, he should have distinguished between energy that is stockpiled and put to use and energy that is fundamentally "cursed" not only in and through bodily excess but in its ability to do "work.")[30] It is not merely a question of our attitude toward expenditure, our "self-consciousness": also fundamental is *how* it is carried out. Waste based on the consumption of fossil or inanimate (nuclear) fuels cannot entail intimacy because it is dependent on the thing *as* thing, it is dependent on the energy reserve, on the stockpiled, planned, and protected self:

> "[This is] what we know from the outside, which is given to us as physical reality (at the limit of the commodity, available without reserve).

> We cannot penetrate the *thing* and its only meaning is its material qualities, appropriated or not for some use *[utilité]*, understood in the productive sense of the term. (*OC*, 7: 126; *AS*, 132; italics Bataille's)

The origin of this destruction is therefore to be found in the maximizing of the efficiency of production; modern, industrialized waste is fundamentally only the most efficient way to eliminate what has been overproduced. Hence the Marshall Plan, proposing a gift-giving on a vast, mechanized scale, is different in kind from, say, a Tlingit potlatch ceremony. "Growth" is the ever-increasing rhythm and quantity of the treatment of matter for some unknown and unknowable human purpose and that matter's subsequent disposal/destruction. One could never "self-consciously" reconnect with intimacy through the affirmation of some form of industrial production-destruction. To see consumer culture as in some way the fulfillment of Bataille's dream of a modern-day potlatch is for this reason a fundamental misreading of *The Accursed Share.*[31] Bataille's critique is always an ethics; it entails the affirmation of a "general economy" in which the particular claims of the closed subjectivity are left behind. The stockpiled self is countered, in Bataille, by the generous and death-bound movement of an Amélie, of a Sadean heroine whose sacrifice puts at risk not only an object, a commodity, but the stability of the "me." To affirm a consumption that, in spite of its seeming delirium of waste, is simply a treatment of matter and wastage of fossil energy in immense quantities, lacking any sense of internal limits *(angoisse)*, and always with a particular and efficacious end in view ("growth," "comfort," "personal satisfaction," "consumer freedom") is to misrepresent the main thrust of Bataille's work. The point, after all, is to enable us to attain a greater "self-consciousness," based on the ability to choose between modes of expenditure. Which entails the greatest intimacy? Certainly not nuclear devastation (1949) or the simple universal depletion of the earth's resources and the wholesale destruction of ecosystems (today).

We face a situation through Bataille, then, in which, to paraphrase the Bible, "the left hand does not know what the right is doing."[32] By affirming the generosity of the self that risks itself, the irony is that, as in 1949, an economy of expenditure—one that affirms the bodily expenditure of sacrifice, of the orgy, of the celebration of cursed matter—will "save the world."[33] Instead of facing—and choosing an alternative to—nuclear war, as Bataille in his day did, today we effectively, and perhaps inadvertently, choose an alternative to ecological disaster brought about by unwise modes of consumption *(consommation).* Expenditure is double, and just as the affirmation of giving, according to Bataille, could head off nuclear apocalypse, so too today we can envisage a model of expenditure that, involving not the expenditure of a standing reserve of eighty million barrels a day of oil, but the wastage of human effort and time, will transform the cities of the world, already facing imminent fossil fuel depletion (what I call postsustainability). What indeed would a city be like whose chief mode of expenditure entailed not the burning of fossil fuel but the movement of bodies in transport, in ecstasy, in despair?

We have no model of such a city. It is up to us to imagine it, to practice it. Of course there have been cities in the past, built around religious expenditure, sacrifice; Bataille examines them in *The Accursed Share.* But modern cities devoted to a cult of the death of God? Nor do we have a model of a monotheist religion that tries to propose a godhead who affirms the unconditioned, the void of his own death. (Perhaps no such model is literally conceivable.) All we have are religions of the Book, religions of useful all-conquering violence, religions that limit expenditure and guarantee the permanence of the self through the worship of a transcendent being.[34] We have seen, so far, energy as inseparable from (sacred) matter and intimacy, the production of and identification with this energy as generosity and risk. We must now think of religion as energy charge: from expenditure as a fixed doctrine, entailing rank, exaltation, and social and personal stability—if not routinized slaughter—to religion as the scattering of doctrine, to sacrifice as the putting in question of the stockpiling of natural resources, of bodies, of victims. Religion, in other words, as the dispersal of any possible codification of the sacred in a Book. And along with this counter-Book, the city as the non-place, the u-topos, of scattering.

Bataille's Religion

The Counter-Book and the Death of God

The Imperative of Religion

The right hand does not always know what the left is doing: generosity in Bataille's universe never can be pure; the gift—of energy, of time, of life itself—is squandered within a grid of opposition, within a social matrix. Truman, presented at the end of the *Accursed Share,* gives, most generously, to a Europe devastated by war: the Marshall Plan. The American president thinks he is giving to further the interests of the United States: a prosperous Europe will buy American goods, and capitalism will be saved. But in his very ignorance, he furthers the development of another economy, one that will save the earth. The new economy of gift-giving will render war obsolete and usher in a new society, a new economy, of the intimate world and its rituals—which is also the economy of the mystic, the "inner experience." The ethics of gift-giving always entails a non-knowledge, an impossibility of willing a specific future as goal or plan. The ethics of expenditure are an ethics of human survival in an economy in which the future—planning, hoping, deferring—is repetitively trumped by the intimate world, by the immanence of squandering in the here and now.

The coming of the new economy will usher in a new relation to excess energy. A cursory reading of *The Accursed Share* might give one the impression of a mechanistic, functionalist view of society based on physiological or zoological models. Just as a given ecosystem can absorb only so many animals of a given species—before the excess is destroyed, in one way or another—so too a given society can only absorb so much energy, so much wealth, before the surplus is destroyed, in a beneficial or harmful way. Usable energy, good for doing "work," is replaced by another, tragic, and savage energy. The power to do evil, if not work. There is, however, a radical difference: people are conscious. They are not merely terms in an organic system with a response typical of all other organic systems. They *know* what is happening, at least in principle.

For Bataille, as we've seen, religious ritual—sacrifice—entails something related to this knowledge, but different from it.[1] Sacrifice is the useless expenditure of excess energy in society. It is burn-off rather than constructive labor; it is the intimate world (or heterogeneous matter) rather than the useful thing. It is action *now* rather than with a later end in view. But sacrifice is quite different from the destruction, say, by wolves of excess deer in a herd. Sacrifice is a way of knowing, consciously, that one does not know—since utility and knowledge (planning, foresight, retrospection) are inseparable. If nature lacks knowledge, if the destruction of surplus deer is a purely natural, unknowing event, then humans, through religion, *knowingly* reestablish contact with a natural realm of expenditure that is closed off from the human world of practical distinctions and coherent knowledge. This is a higher-level or lower-level (as the case may be) *continuity* with nature, a knowing, so to speak, with the unknowing of nature. Such in any event is Bataille's thesis in his work *Theory of Religion.*

Much of Bataille's major, three-volume aphoristic work, *The Summa Atheologica (Inner Experience, Guilty, On Nietzsche),* is concerned with this very problem. The project and representation—and writing—are inseparable. To write is to plan, to project, to put off pleasure, to render permanent, to put to use what is there, what is natural. But this is precisely what religions do: the major religions are all "religions of the Book." Mystics, too, scribble endlessly. And yet if that is the case, how can we say that these religions are inseparable from an event, an experience (which is not an experience)[2] such as sacrifice, which entails a moment of intimacy, of uselessness, one radically foreign to the project, one that establishes contact with a continuity that is logically prior to all self-interested human intervention, to all self-interested writing?

Of course, the first gesture can be to reject any established religion, and this is what Bataille does (see the prewar essay "The Psychological Structure of Fascism," for example).[3] Religion in this view is the reappropriation of sacrifice and of the cursed matter associated with it by a cultural practice of utility: God, in other words, becomes the principle of utility. He can be prayed to, and he will help us, reassure us; rather than hideous, unassimilable matter (the left-hand [impure] sacred), buzzing with the energy of Sadean atoms, he is pure, elevated, eternal, unchanging, omniscient.

This is where the *Summa Atheologica* comes in: this entire, maddening, exhaustive, exhausting work, composed of hundreds of pages of aphorisms, miniessays, diary notes, and fragments, has as its goal one central question: how can one *write* an inner experience, in other words a religious event that in a sense doubles the sacred of sacrifice *before* it was made pure, holy, eternal, good, and useful? If we reject established religion as useful, fixed, predictable, predicting, we are left with a sacrificial moment, individual or collective, that, it would seem, cannot be codified, cannot be taught, transmitted, used, put in words. An unthinking moment, ecstatic, perhaps, but as averbal as the death agony of the deer. But—and this is then the giant question—how can it be written? Wouldn't trying to write it, as Bataille does, end up making it permanent, rendering it a teaching, a body of knowledge—useful, in spite of itself?

At times we are led by Bataille to think of his inner experience as somehow congruent with that of the mystics—especially when we learn that he has been dabbling in yoga and other practices. But, true to form, he rejects traditional mysticism because it attributes itself to a divine intervention; for this reason the only mystics who seem to interest him are figures such as St. Angela of Foligno, who (it seems) denied the existence of God.[4] Bataille's mysticism is atheistic because the radicality of the experience itself precludes the possibility of its authorization by a stable and knowable Godhead. The dark night of the event precisely cannot be known, and it certainly cannot be ultimately explained (away) or attained through recourse to a project operating under the aegis of a coherent deity.

A radically atheistic mystical experience (if that is what one could, for the moment, call it), however, still poses problems. How is one to communicate it? And if it is not an experience, how can one even conceive it, let alone write it? Put another way, how is one to avoid rendering the experience as just another, well, *experience:* another version of what all religions do (namely, the rendering of the unconditioned, the sovereign, as a fixed meaning that serves a purpose)?

This is, so to speak, the Scylla and Charybdis of Bataille's experience: either it is writable, in which case it becomes just another positive event, another element of certifiable religious experience or doctrine, or it avoids the latter, negates it, in which case we are at best back to the natural expenditure of the sun or of any ecosystem that regulates itself through burn-off. In that event nothing at all could be said on the basis of the experience itself; all we could formulate would be scientific propositions (the temperature of the sun, the deer population before and after, etc.).[5]

To maintain this sort of either-or, however, and to hold it against Bataille, is to misinterpret Bataille's project. Perhaps Bataille himself contributes to this misunderstanding by peremptorily dismissing established religions because they turn the radicality of the sacrificial event into a merely useful element in their liturgies. Here of course the Catholic Church is the prime exemplar, since the consecration carried out in the Mass is literally the sacrifice of Jesus on the cross but converted into a reassuring ritual that gives meaning to our lives. But a more careful reading of the *Summa Atheologica* reveals that, far from dismissing language and ritual, Bataille puts it to work, so to speak, against doctrine and against codified mysticism. His goal is not simply to dismiss those things, because to do so would be merely to establish his own ineffable doctrine, his own mysticism, in their place. Rather, recalling the model presented in *Theory of Religion,* where man's reconnection to the continuity of nature takes place through and against that which is most opposed to nature—the codified ritual of sacrifice, linked to human language and individual identity (consciousness)—in the *Summa Atheologica* Bataille ultimately presents a writing that establishes a knowledge of what cannot be known, of what is most resistant to any formulation of knowledge, to any writing.[6] Compared to established religion, his knowledge recognizes its impossibility, its loss, in its attempt to write any mystical knowledge, and thereby communicates (in a Bataillean sense) with, opens itself out to, the loss of sense. Sense is produced only in order to be squandered, slipped into the void, just as wealth is produced only to be given and destroyed. Sense is self-reflexive, "consciousness," in that it is nothing other than the awareness of its own movement into non-knowledge.

Bataille's tactic is therefore not so much to formulate a new doctrine or to escape all doctrine—let alone to escape writing as such—as it is to rewrite existing doctrine, existing books, from the perspective of a writing of the impossibility of writing (the impossibility of formulating a higher, eternal doctrine, communicating a union with a supreme Godhead, etc.).

Bataille in fact confronts not one but three religions of the Book, three doctrines that base their truth on a kind of immutable, sacred writing. In each case, Bataille does not so much refute or deny the other's book so much as he rewrites it, breaks it open, subverts it while affirming it. It is a deconstructive tactic *avant la lettre:* what counts here is the gesture of establishing a doubled religion, a doubled Book, which carries out exactly the same procedure we saw, in chapter 2, in the study of economics, history, and politics. Here too, as in the case of economics, a Bataillean religion entails a limit of meaning, coherence, knowledge, which is crossed, transgressed, by a continuity of not-knowing, of extinction of the self and of all projects. Out of this duality arises an *angoisse,* a dread before the nameless night that both opens the possibility of, and defeats, any religious certainty, any textual doctrine or mystical certitude.

Thus questions of economy, energy, and expenditure in Bataille are from the first religious questions as well. But we might say that this is a strange religion in that it entails a counter-Book that writes the impossibility of writing, of codification, of religion. In this the counter-Book is of the greatest importance because it is the Book that always serves as the temptation of a certain literalism: the idea that not only is there a transcendent God who serves as a human ideal (he is just, permanent, transcendent, vengeful, peaceful, etc.), but that God's Word is definitive, not subject to the vicissitudes of interpretation. The religion of the Book presupposes not only a fixed doctrine, absolute and unchanging, but a corps of specialists (clerics) who teach it as Law and impose it on all believers. One could argue that behind each of the religions of the Book there is a perfectly coherent position that demands complete devotion to a definitive truth. Bataille's stress on the writing of a counter-Book, a text that opens out, doubles, and loses any unitary doctrine, can only be seen as a direct blow aimed at any attempt at a repressive and totalitarian religious doctrine—a blow elaborated from within, but against, the imperative of religion. Non-religion, we might call it, at the limits of religion: the religion of the fall of God.

For Bataille two secular religions—those of Sade and the Hegelian Alexandre Kojève—were, in the modern era, every bit as powerful, if not more powerful, than those of the more traditional monotheistic cults. They threatened to reduce their believers to ciphers, ditto-heads saturated with an all-powerful yet strangely indifferent doctrine. Both Sade and Kojève saw their projects as in some sense replacing religion with something more absolute than God: the text of the Book itself and finally the sheer material presence of the Book. After God, in the space of his absence or demise, there remained the ultimately cold, cruel Sadean Book and the definitive Book of Hegelian-Kojèvian absolute knowledge. In that sense, Sade and

Kojève's works are superreligions, and Bataille's rewritings of these two authors constitute the grounding for his "atheology"; in a work devoted to the event of the death of God, he is in fact more concerned with these seemingly atheistic authors than he is with the traditional religions they seem to have displaced. Nevertheless, Bataille's focus on the death of God also entails an affirmation of the Judeo-Christian God *as dead.* For this reason Bataille must also return to the God of Genesis, not only to affirm the resistance of God to any atheism that would "go beyond" him (such as the atheism of Sade and Kojève)—the resistance of God to any modernity, in other words—but to reaffirm through his critique of Genesis a counter-Book of the death of God: the *Summa Atheologica.*

Bataille and the Sadean Book

The first book Bataille rewrites and transgresses is that of Sade. We've already seen Bataille's use and abuse of Sadean matter, and eventually of Sadean politics. But Sade can also be seen, and read, as a book, every bit as untouchable as the Bible. Sade is, as Pierre Klossowski (in *Sade mon prochain,* 1967) noted many years ago, essentially a religious author. As we have seen, Sade's God is a fiction, disavowed, yet he is incessantly re-created, in his own way as necessary as the Christian God.

The entire Sadean corpus has a kind of absolute existence, not so much because it teaches or because through its teaching it enacts the defilement of God, but because it offers an absolute, and absolutely natural, substitute for religious experience. Nevertheless it is built on paradox: one hates God because he is false, is a mere projection of the weak; but by defiling him as weak, as a fiction, one achieves one's greatest orgasms. The Sadean hero is caught in a diabolical fiction in which what is disavowed must constantly be reestablished—all the while in the recognition that its reestablishment is mere fiction.

One solution to this double bind is to establish something else more absolute, more real, than God to serve as a constant referent while God shimmers in and out in fiction. That absolute is the Sadean Book, itself dependent on the basic premise that God is to be abused as a fiction. In that sense Sade's Book is the ultimate truth, not only replacing religion but affirming an energy that goes beyond nature in its absolute power.[7] Like the latter, the Sadean experience of nature is overarching, total: it is seen to replace, entirely, the (false) promise of religious experience, and it is available (as in the case of religion) through the repeated, indeed compulsive, perusal of the book.

It can be said that Sade's pornography is the logical consequence of the materialist theory of, among others, Condillac and Helvétius. Sensations are the measure of truth; sensations are processed and retained through the imagination; in this way the sensations can be re-created. Caroline Warman, in her excellent book on Sade's materialism, stresses that for Helvétius the beautiful is nothing more than the creation of the strongest sensations; ultimately "a painting is measured as a sensation, and so is a text" (2002, 53). "The art of writing consists in the art of exciting sensations," adds Helvétius (cited in ibid.).

Language is thus a privileged sense-perception because it "awaken[s] the imagination" and causes previously experienced sensations to be relived (Warman 2002, 53). If, moreover—as Helvétius argued—the strongest sensations are the most beautiful, then, for Sade, his pornography will constitute the pinnacle of beauty, since the most powerful and hence most pleasurable sensation of all, murder, will be relived in the imagination and then reenacted sensually through the reading of his books.

But pornographic pleasure will involve more than beauty, since Sade's Book will also have a privileged access to the ultimate origin point of truth: Nature. If Nature rewards the libertine for actions that advance and accelerate her "transmutations," then the Book, in causing a mimetic doubling (in the reader) of the sensations of the murderer, will bear, among all books, a privileged relation to the truth of Nature. Moreover, the Sadean Book will not only embody beauty through the rewarding stimulation of the imagination and senses, it will also convey the truth of Nature in the doctrine of Sadean crime. Thus the Book is both ultimate beauty and truth: beauty in the concentration, in aesthetic form, of the most rewarding and genuine, sensual, natural, experience (murder); and truth, since the pleasure of murder is the highest good of Nature and the Book contains its irrefutable doctrine. The Book does not simply embody the doctrine (of pleasure) or convey it; it explicitly reflects upon it, formulates it, and this formulation is itself inseparable from pleasure (the imagination cannot cause sensual experience to be relived without a conceptual grounding—else the sensual experience would be nothing more than animal sensation, and not the privileged grasping, and thus privileged intensification, of Nature).

Sade's pornography is the greatest work, supreme truth and beauty, and all other work is either subsidiary (other literature, for example) or in error. Not surprisingly, however, there is another side to Sade's Book, one that tends to be ignored by Sadean commentators. It too entails the ultimate power and authority of the Book, but this time one that does not ennoble the reader and bring him or her to the highest level of Nature herself, both understanding and intimately participating. This time the reader is cast down, debased, abased, left a shattered ruin. In other words, a victim.

The model presented up to this point connects reader, narrator, and Nature; we participate on the most intimate level in Nature's works, and she rewards us; we murder, she gives us the greatest pleasure. Missing here is the fact that there is always a victim, a double; our senses are the most stimulated by the identification with the suffering of another, weaker, person. If the vicious always triumph, it is the weak, the virtuous, who are executed (not for their guilt, but for their innocence): God, the most virtuous (and, unfortunately, the most fictional), first of all. And yet there is an obvious identification, a communication between victim and murderer: the maximum stimulation of the victim's senses, to the point of the death agony, causes (and alone causes) the maximum stimulation of the criminal's senses. The libertine identifies with the "other," the always-already murdered, in the very act of negating that other; this alone stimulates the libertine and makes possible the greatest identification of all, with Nature (the ultimate author of all murder).

Now at this point the tables turn, and the reader comes to identify no longer with the killer, but with the killed. The reader, in other words, becomes Sade's victim—not, of course, by being killed, but by being morally destroyed, made weak, made a potential victim. Any reader of Sade is familiar with the drill: one sees it most clearly in *Juliette,* demonstrated over and over again. First comes an orgy, with various titillating details: all the varieties of coupling that caused later commentators to see in Sade a precursor of Krafft-Ebing or Kinsey. And there is a kind of easygoing irony; nouns ("indecency," "horror," "monsters") are overturned in their meaning and become positive markers. It is all fairly harmless; all the characters come, none (so far) die. The reader is brought in, seduced, by a pornographic fiction not that different (in the eighteenth century or now) from thousands of others. But then the strategy is intensified: from a free-wheeling, even humorous orgy, suddenly someone—Clairwil, Olympe Borghese, Juliette herself—proposes to turn up the volume a bit. And soon the victims' bodies accumulate, sex scenes become torture scenes, and the reader, still sort of enjoying it all, starts to feel . . . guilty.[8] Ashamed. Recognizing the repulsive and fundamentally immoral nature of his or her desire. But all the while still reading, compulsively.

Unlike the other materialists of the period, who wrote as *philosophes* and who, like all *philosophes,* flattered the reader by assuming his or her intelligence, Sade, by intensifying only slightly the materialist conception of beauty—in pornography—degrades the reader and brings him or her to the point of being ashamed by that to which she or he has acceded. In a word, one feels like a disgusting creep when one is aroused by such violent and loathsome trash. And that is the Book's (im)moral power: the virtuous person (anyone, in other words) is targeted in his or her virtue, marked as a victim, then left to rot. Like any victim, the reader, a standing reserve of goodness (at least in his or her self-estimation), is fucked, used up (morally at least), and dumped like refuse. Virtue is the weak spot that allows guilt to reside in the victim and fester; one recognizes, through one's very arousal, through one's endless reading, an irremediable identification with the most horrible pervert, and one is powerless to do anything about it. The Sadean reader is for this reason a hypocrite, a liar, a scoundrel, but a weak one. The problem is that one can never be as wicked, as powerful, as a Sadean hero (one always retains a residue of virtue, of political correctness, of whatever), and in this impossibility resides the curse of Sade's Book, its vampiric destruction of its faithful and never evil-enough reader.

Still, the Sadean implied reader goes on, imagining her- or himself virtuous, or at least correct (if only in a political sense), while continuing to get off on fantasies of crime—not just of simple murder, but of all sorts of betrayal, theft, blackmail, fraud—the destruction of elevated personages, families, cities, whole nations and religions. Bad faith, on our part, on the part of the reader, resonates in horror; it can no longer be ignored. The Book has transformed itself, first moving from abject pornography to absolute beauty (practical beauty through the stimulation of the senses), and then from true beauty to purveyor of truth—the truth not only of Nature, and of crime in Nature, but of the abasement and hypocrisy of its reader, of all readers. The reader can never hope to attain the cold criminality of the Book, its cynical manipulation and certainty of our own weakness, and in this the Book reaches a pinnacle of inhuman superiority. One thinks again of the apathy that characterized, according to Blanchot, the Sadean hero: now that apathy characterizes the Book itself, absolute, sublime in its detachment, its sheer materiality, its elevation above the weakness and falseness of the human will. That all-powerful apathy radiates from the Book, glorying in its materiality, nothing more than simple matter, paper and binding.

Now I would argue that the *Summa Atheologica,* like all of Bataille, repeats Sade, but also inverts him. In Bataille it is not the reader who internalizes the mortality of the victim; it is the narrator, the "author" himself. As we have seen, Sadean selfishness in Bataille becomes a kind of profound generosity. Amélie, Bataille's favorite Sadean heroine, longs to give herself, sacrifice her own life, to further a criminal's commission of a stupendous crime. In the *Summa* Bataille does not so much represent criminal heroes as he represents himself, like Amélie, as a reader of violence and a willing victim: we move from the depiction of violent death for the reader's delectation and degradation, to the depiction of the writer in the act of contemplating (his own) violent death as an atheological event. This is an important difference, for now we are directly considering the larger ethical—or spiritual—question of the avid contemplation, or perception, of murder, of one's own (affirmed) murder. This is the very question that Sade, in his profound apathy, did not bother to pose.

Specifically, Bataille contemplates a photograph of the torture and execution of a Chinese man, apparently a rebel during the Boxer rebellion. The man is slowly dismembered, in successive photographs, and, "in the end, the patient *[sic],* his chest stripped of skin, twisted, his arms and legs cut off at the elbows and knees. His hair standing straight on his head, hideous, haggard, striped with blood, as beautiful as a wasp" (*OC,* 5: 139; IE, 119).

This kind of contemplation has its obvious precursors both in the reading of Sade and in the contemplation carried out by Jesuits in training. The disciple of St. Ignatius of Loyola, Bataille reminds us, imagines in the most graphic way the wounds and suffering of Christ in order to, according to Bataille, "attain the non-discursive experience" (*OC,* 5: 139; *IE,* 119). In other words, by "focusing on a point," the novice attains a Christ-like receptivity, but does so only starting from words. Drama—the drama of Christ's suffering and death—serves to stimulate the imagination, but, according to Bataille at least, the dramatic (and narrative) identification only takes place thanks to discourse—the vivid retelling of the story of the Passion.[9] The aspirant Jesuit might leave words behind, but he only does so thanks to a coherent discourse that he follows.

In Bataille's case, on the other hand, "the full communication that is the experience tending to the 'extreme' is accessible to the extent that existence is successively denuded of its middle terms: of what proceeds from discourse, and then, if the mind *[esprit]* enters into a non-discursive interiority, of all that returns to discourse, due to the fact that one can have a distinct knowledge of it" (*OC,* 5: 135; *IE,* 116).

The "ecstasy before a point" (*OC,* 5: 133; *IE,* 114), then, entails the preliminary loss of language; it is not the contemplation of a given agony through a story but instead a kind of visual identification with a senseless agony. Thus "drama" in Bataille's sense connotes immediate identification and projection rather than a progressive discursive contemplation that eventually leads to a larger, nondiscursive, spiritual identification. For if the movement continues to depend on a narrative mediation, the final point of identification—God—will himself be dependent on, and a function of, discursivity, labor—the project.

The immediate identification with violent death takes us out of Loyola's orbit and into Sade's. The Sadean libertine gets more powerful, and certainly more discursive, as he or she contemplates (and enjoys) the violent death of the victim. Imagination, language, and physical stimulation are inseparable, both on the part of the (infinitely powerful) fictional subject within the story and on the part of the (infinitely weak) implied reader perusing it. This moral transfer, the draining of energy from victim (the reader) to victimizer (the Sadean Book) takes place through drama and projection—the same devices used by Loyola. Contemplation, story, and deliverance—or ignominious destruction—go hand in hand.

In Bataille, on the other hand, there is no imagination, at least not in the sense in which it contributes to the power, stimulation, or degradation of an isolated self. On the contrary: for Bataille the self is an *ipse,* an identity characterized not by its constitution as a stable selfhood, but instead as the movement of *loss* of identity.[10] This *ipse concentrates* on the other, the representation of being, a "point," in agony ("this point can radiate arms, scream, break out in flames" [*OC,* 5: 137; *IE,* 118]). Further, "It's in such a concentration . . . that the existence has the leisure to perceive, in the form of an inner brilliance *[éclat],* 'what it is,' the movement of painful communication that it is, which goes no less from inside to outside, than from outside to inside. And no doubt it's a question of an arbitrary projection" (*OC,* 5: 138; *IE,* 118).

The "ecstasy before the point," which is the inner experience, entails a kind of violent, nonmediated projection of a self that is *not* a coherent self (the *ipse*) into another, a double, which in turn projects back onto the *ipse* its violent dissolution. This is the movement of "communication," as Bataille calls it, not a communication of information but instead an "incessant gliding of everything to nothingness. If you like, time" (*OC,* 5: 137; *IE,* 118), the immediate slippage of the *ipse* into the other *ipse* in an identical state of loss—and into "the dereason *(la déraison)* of everything" (*OC,* 5: 134; *IE,* 115).

Impossible as this "negative mysticism" is to talk about—and this is precisely Bataille's point—two things stand out, which both put in question the total certainty and total efficacity of Sade's Book.

First, the *identification* with the other. This can certainly take the form of cruelty, of betrayal, but the cruelty goes in both directions—from inside to outside and vice versa. As Bataille puts it,

> The young and seductive Chinese man I've spoken of, given over to the labor of the torturer—I loved him with a love in which the Sadistic instinct had no part: he communicated to me his pain or rather the excess of his pain and that was precisely what I was looking for, not to enjoy it *[pour en jouir],* but to ruin in myself what was opposed to ruin. (*OC,* 5: 140; *IE,* 120).

The contemplation of the man's torture and death is not a pleasurable identification that in some way reaffirms one's own power and pleasure, nor does it entail the reader's straightforward moral degradation; on the contrary, it opens the way to an "experience of ruin" that cannot be an experience because the latter term implies something gone through that can be remembered, described, understood through discourse—whereas the opposite is the case here.

Moreover, writing of a contemplation of violent events, volcanoes, fire in the sky, overwhelming destruction, Bataille stresses that "the night" nevertheless surpasses the straightforward violence because, in the night, "there is nothing sensory *[il n'est rien de sensible]*" (*OC,* 5: 145; *IE,* 125). The point is not that the night is spiritual, but rather that, in opposition to Sade, contemplation of the night, entering into a profound communication with it, does not entail or result from a sensory experience—one that can be measured, analyzed, or known scientifically.

Sade's reason, and Sade's Book as well, are broken apart by this model of violent "meditation," which nevertheless shares with Sade a willing contemplation of horrible violence. Sade's Book and the self of the libertine hero are shattered, scattered, in a "communication" that is not discursive, that does not lead to coherent discourse, is not sensory, and is not, in the conventional sense of the word, sadistic. If the reader is shattered, it is not because the Book has willed it while staying immune from it; on the contrary, the Book is shattered every bit as much as the reader.

This is another way of saying that the reader is not somehow outside Bataille's Book, acted on by it; on the contrary, the reader inhabits the Book, occupying its writing practice, its "narrator." Just as subject and object switch poles and are lost in dissolution ("communication"), so too in Bataille's model of reading the reader inhabits him, speaks or writes him, in his violent "experience."

> The *third,* the companion, the reader who moves me *[m'agit],* is discourse. Or, better: the reader is discourse, it is he who speaks in me, who maintains in me the living discourse addressed to him. And, no doubt, discourse is a project, but it is still more this *other,* the reader, who loves me and who already forgets me (kills me), without whose present insistence I could do nothing, and would have no inner experience. (*OC,* 5: 75; *IE,* 60–61; italics Bataille's)

Bataille projects the reader as the very discourse "in" him that tracks and communicates an untrackable and incommunicable "experience." Sade's Book's certainty, its dominion, its (im)moral sublimity that goes beyond Nature's own inhuman sublimity, is here recast as a discourse, and finally a Book, that is riven with a self-destructive project in which the impossible "communication" of the "experience" is inseparable—is indeed the same as—an experience that we could take, at first in any case, as a strictly personal or private event. In this sense Bataille's Book is not only not a coherent, absolute production of sense, it is not even a Book at all, since it communicates nothing that is not already the reader's discourse and failure of discourse. But Bataille's Book, despite its form, is certainly not a journal or memoir in the conventional sense. Bataille confronts Sade with a complex movement, writing what cannot be written through a personal statement that is not personal, communicating to the other an "experience" that is not an experience, that which is the other's already, communicating communication in a fractured language that cannot communicate sense, that is not discourse, communicating communication as dissolution, both of the "narrator" and of the "reader" (roles that are themselves necessarily arbitrary and deceptive).

Rather the Book is a hole, in Bataille's well-known formulation: "I write for he who, entering my Book, will fall into it as into a hole, and won't get out" (*OC*, 5: 135; *IE*, 116). The reader falls, but necessarily Bataille does as well, alongside; indeed, the Book, the writing, the discourse of the reader as Bataille, and Bataille as the reader (who kills him), is a hole that has fallen into itself as into a hole.

All this still leaves a question, and one that can best be addressed through a consideration of Bataille's undoing of another absolute Book: that of Hegel (the *Phenomenolgy*), as reinterpreted by Alexandre Kojève. What is this "night" into which the *ipse* is projected, the "night" that seems to lurk behind and through any "communication" with the violence of another *ipse*? Is this "night" simply an impersonal event, like Blanchot's *neutre,* or Levinas's *il y a,* a "rustling" of radically formless, primordial matter?[11]

Certainly what Bataille will come to associate with this "night"—what he calls "le non-savoir," non-knowledge—bears certain similarities to a Sadean matter that is not so much a scientifically calculable one derived from Helvétius or d'Holbach (one that entails a simple calculus of pleasure and pain) as it is one of a monstrous energy, a profound "transmutation," harkening back beyond Sade to the alchemical tradition of Bruno. An energy that, like Bruno's God, is logically prior to the establishment of systems of coherent oppositions (it is not simply "not useful," but it is instead somehow beyond or on this side of usefulness: opening the possibility of usefulness, of work, without submitting to it). Yet this "night," shuddering with volcanoes under a fiery sky—with the heterogeneous (or base) matter matrix of incessant production-destruction—is also inseparable from the scattering of the most perfect philosophical *work:* the Hegelian-Kojèvian Book. This Book, like that of Sade's *philosophie,* is the definitive embodiment of a certain immutable matter and of a knowledge so absolute that it is no longer even human, no longer, strictly speaking, alive. It is Bataille's repetition of this Book that will plunge it back, incessantly, into the night of pulsing heterogeneous matter and the ruin of discourse.

Bataille and Kojève's Book

Alexandre Kojève, the master interpreter of Hegel's *Phenomenology,* held sway over much of the independent-left intelligentsia of the 1930s and 1940s. As is well known, many of the era's leading thinkers—Bataille,

Raymond Queneau, Maurice Merleau-Ponty, Jacques Lacan—religiously followed Kojève's lectures at the École pratique des hautes études en sciences sociales. Beyond this select group, others, such as Jean-Paul Sartre and Maurice Blanchot, were heavily influenced by the "Hegelian wave" of thinking, even if they did not show up every week chez Kojève.[12]

What Kojève provided the non-Stalinist left was a model of Hegel that in a sense replaced Marxist economic determinism with a thinking of the close connections between liberation, human self-recognition, labor, and philosophical truth. For Kojève, the end of history will arrive (or has already arrived) in and through a State that makes possible (and necessary) the mutual recognition of the freedom of each by all others. To be Human, in Kojève's version of Hegel, is to labor; the movement toward human liberation is the negation of existing reality, existing Nature, and the creation of the State that institutionalizes and makes permanent this liberation. Time is a function of the negating labor of material and philosophical activity. At the End, this philosophical labor will be over; the subject/object opposition will end, Man will be free, Time will be over, and the universal, homogeneous State will guarantee the permanent recognition of Man by Man.

Writing in perhaps a more Heideggerian than Hegelian mode (but at the same time, in an anthropological mode Heidegger would have found quite bizarre), Kojève stresses the importance of death in this process.[13] Freedom, in effect, is the application of finitude; my negating and destroying one reality produces another; freedom entails my recognition of my own finitude in the very act of externalizing it; it entails as well my recognition of the freedom or desire of the other as the application of her or his own finitude in the negation and transformation of the world. And finally—perhaps the most controversial point in all of Kojève—this finitude is not only what characterizes me, i.e., my action in the world; it characterizes the very movement of History and human Time. If philosophy and the transformation of society is a kind of labor, this labor, like all activity, is finite: it ends. Man externalizes death, the finite, in labor, and in the labor of philosophy.[14] Like all work, philosophy, the truth of the movement of human Spirit and labor in the world, ends, dies, and Man as the author of this movement dies as well. At the End of History, the final recognition, the attainment of Absolute Knowledge, ends in a State that guarantees this Knowledge, embodies it. After the End, so to speak, there can be no more philosophy, since definitive Truth has been found. Nor can there be any more political evolution or revolution, since the definitive State has been attained. All meaningful change—all transforming labor having to do with society, religion, and philosophy—is over.

Such a vision had enormous appeal in interwar Paris; here was a model that considered human liberation not just in terms of recovered surplus value—of Marxist economic considerations—but in terms of the grand march of philosophy. Kojève, after all, awarded pride of place to the Wise Man *(le Sage),* the one who formulated the Truth at the End; he it was who wrote the Book, the definitive doctrine of Being, finitude, labor, History, and Death. Such a Wise Man was not to be considered an intellectual; on the contrary, while the intellectual was considered a dreamer, one who promoted entirely subjective and partial "truth," the Wise Man wrote and published—disseminated for all—what was the definitive Truth. The Wise Man was, in other words, a kind of the definitive (and final) thinker who went beyond the obvious vices of the traditional French intellectual. Such a vision could only have an enormous appeal in Paris, where, by the 1930s at the very latest, in a period of massive cultural meltdown, there was considerable disenchantment with the evident weakness represented by the traditional figure of the intellectual (no rigorous philosophical analysis; all scintillating talk and no action; too much emphasis on aesthetics and novel writing, not enough on dialectics).

Kojève's version of Hegel, however, had great appeal for another reason: the seemingly irrefutable demonstration of the End of History and of Man resonated not only with Marxism, giving its promise of definitive human liberation an apparently solid philosophical grounding; it resonated as well with avant-garde thought. The surrealists too had posed their version of liberation as coming after the advent of the classless society foreseen by Marxism; now, in Kojève, they could imagine the exotic activities Man could engage in *after* his death. With all labor definitively over, in other words, the surrealist dream, many thought, could come to the fore: all the activity that could *not* be limited to logical activity geared to a practical goal could finally flourish. Among others, Raymond Queneau, a former surrealist, considered this question extensively in his writings, such as his novel *Le Dimanche de la vie* (1951, *The Sunday of Life*)—a Sunday in which, no doubt, all useful labor having been completed, Man, or post-Man, would be free to rest, or better yet to engage in all sorts of (by definition) nonproductive activity, not least of which would be play, eroticism, aesthetic and athletic activity: anything, in other words, but the work held so dear by the hard-line Marxists of the French Communist party.

What is interesting for our concerns is that Kojève himself was very much interested in the problem of what comes *after* the end, as is made evident by several well-known footnotes added to his *Introduction* when it was published.[15] These concerns are, it turns out, inseparable from his thinking of the Book—the definitive work by Hegel as amended by Kojève himself—that would serve as not only the codification of the movement of Truth and History, but as the final and definitive resting place of Spirit.

Having emphasized to such a great extent, following Heidegger, the role of finitude and the recognition of Death, Kojève was hard put to imagine how, exactly, Man would die. Kojève in fact uses the Book as a kind of repository of Man; Man, dead, nevertheless continues to live so long as he reads the Book.

> Certainly, [at and after the end of history] the Book must be read and understood by men, in order to be a Book, in other words something other than paper. But the man who reads it no longer creates anything and he no longer changes himself: he is therefore no longer Time with the primacy of the Future or History; in other words he is not Man in the strong sense of the word. The man is, himself, a quasi-natural or cyclical being: he is a reasonable animal, who changes and reproduces himself while remaining eternally identical to himself. And it is this "reasonable animal" who is the "Absoluter Geist," become Spirit or completed-and-perfect *[achevé et parfait]*; in other words, dead. (*ILH,* 394)

Here we learn that the "death of Man" is not his physical extinction but is rather his survival in a minor mode. He may not be "Man in the strong sense of the word," but he is still man (lowercase), alive, reading. Or somebody is still reading, somewhere, still recognizing the Truth. Man was the "Concept or Logos that reveals Being" (*ILH,* 416), but now, since Man is dead, the concept, now Being become Spirit, resides in the Book, definitively written, completed: "It is necessary to see how the existence of the Spirit, in the form of the Book, differs from its existence in the form of Man" (417). It differs, of course, in that Man was in constant flux, in constant Historical transformation. Now that History is over, the Book, never changing, fully objective, is Absolute Knowledge: "It is not Man, it is not the Wise Man in flesh and blood, it is the Book that is the appearance *(Erscheinung)* of Science in the World, since this being is Absolute Knowledge" (384).

It is not enough for the Book merely to exist, however; objective as it is, at the same time (apparently) it can exist only as Spirit if it is read. This is how it becomes objective.

> To turn out to be true, philosophy must be universally recognized, in other words recognized finally by the universal and homogeneous State. The empirical-existence *(Dasein)* of Science—is thus not the private thought of the Wise Man, but his words *[sa parole]* universally recognized. And it is obvious that this "recognition" can only be obtained through the publication of a book. And by existing in the form of a book, Science is effectively detached from its author, in other words from the Wise Man or from Man *[du Sage ou de l'Homme]*. (*ILH,* 414)

Man, then, presumably does continue to exist, if not in the "strong sense of the word." He is a citizen of the State, and he, or someone for him, reads the book. This does not mean, of course, that "reading" is or will be anything like what we understand by that term: since History is completed, nothing new can happen, there will only be one definitive understanding of the book. Indeed by publishing it the book will be subject to the "dangers of being changed and perverted" (*ILH,* 414 footnote). That would be bad reading; good reading, on the other hand, is therefore sheer repetition done either by or under the aegis of the universal State.

> In effect, the content of Science only relates to itself: the Book is its own content. Now, the content of the Book is only fully revealed at the end of the Book. But since this content is the Book itself, the response given at the end to the question of knowing what is the content can be nothing other than the *totality* of the Book. Thus, having arrived at the end it is necessary to *reread* (or rethink) the Book: and this cycle is repeated *eternally.* (*ILH,* 393; italics Kojève's)

So far so good. Man lives, sort of. Reading continues, if we can call sheer repetition reading. But there is a problem. We know that Man is dead, even though he keeps reading. Is the book dead? Is Spirit dead?

At first, we would have to say "no." Spirit is definitive knowledge at the end: it is what is known for good, Spirit recognizing itself, so to speak. The

Book as Spirit in that sense is the truth of God and religion: Although Kojève does not use the term, we would have to conclude that Spirit is eternal life—if life can be seen as unchanging, forever the same (not the simple eternity of an individual soul). If real Spirit, "Eternity," is revealed through Time, reaching its plenitude at the end of Time (the End of History), then God, who is purely atemporal, will be Spirit only until Time, and History, are fully recognized as what they are: in the Book, at the End.

> As long as Time lasts, in other words until the arrival of Science, Spirit is revealed to Man in the form of a theo-logical Knowledge. As long as History lasts, there is thus necessarily Religion, and, if you like, God. But the final cause and the profound reason for the existence of Religion (and of God) is implied in the very nature of Time or of History, in other words, of Man. (*ILH,* 389–90)

The Religions of the Book—the monotheistic religions, Judaism, Christianity, Islam—are thus replaced, at the close of History, by the Book itself, which is the definitive account of their supersession. Kojève does not say this, but it would be quite logical to conclude that the prayer and meditative practices associated with these religions are replaced by reading. But not just any reading: reading that remains absolutely faithful to the Book, reading that does not interpret, question, or misunderstand: reading that repeats, endlessly, Spirit. Spirit is now nothing more than this total reading and the physical Book.

Spirit is life, or what remains of it. But is it death as well? At the end, with the triumph of Spirit as Knowledge and Sense in repetitious reading, there is also death: of Man, and of History. If the Book is no longer part of Man (since Man is dead), then what will be the status of the Book? How can it be alive?

> The fact that at the end of Time the Word-concept (Logos) is *detached* from Man and exists—empirically no longer in the form of a human-reality, but as a Book—this fact reveals the *essential finitude* of Man. It's not only a given man who dies: Man dies as such. The end of History is the *death* of Man properly speaking. There remains after this death: (1) living bodies with a human form, but deprived of Spirit, in other words of Time or creative power; (2) a Spirit which exists—empirically, but in the form of an inorganic reality, not living: as a Book which, not even having an animal life, no longer has anything

> to do with Time. The relation between the Wise Man and his Book is thus rigorously analogous to that of Man and his *death.* My death is certainly mine; it is not the death of an other. But it is mine only in the future: for one can *say:* "I am going to die," but not: "I am dead." It is the same for the Book. It is my work *[mon œuvre],* and not that of an other; and in it is a question of me and not of anything else. But I am only in the Book, I am only this Book to the extent that I write and publish it, in other words to the extent that it is still a future (or a project). Once the Book is published it is detached from me. It ceases to be me, just as my body ceases to be mine after my death. Death is just as impersonal and eternal, in other words inhuman, as Spirit is impersonal, eternal and inhuman when realized in and by the Book. (*ILH,* 387–88 footnote; italics Kojève's)

Kojève wants to distance the Book from the intellectual's product, which is subjective and has meaning only as his fantasy—it is so personal that it is mere wit, and cannot be made objective and communicated as such (see *ILH,* 403). On the other hand, living Spirit implies that Spirit might change, develop—which is impossible. So we have a metaphor: the Book as cadaver. Kojève does not come out and say the Book is dead, but its separation from its author, as the body is separated from the self, certainly leads to this conclusion.

And yet, at the same time, if the Book is not alive, perhaps reading is. But note that reading is no longer a labor, an act of negation and construction. It is not temporal in the conventional sense: how can there be an atemporal reading?

> While changing, the Book remains therefore *identical* to itself. The Time in which it exists *[dure]* is thus natural or cosmic, but not historical or human. Certainly the Book, in order to be a Book and not bound and stained paper, must be read and understood by men. And if, to read the Book, Man must *live,* in other words be born, grow, and die, his life is in essence reducible to this reading. (*ILH,* 385; italics Kojève's)

Time is reading; the Book is its reading; reading is understanding; life is this reading. Man, who is dead, nevertheless lives through the eternally unchanging reading of a Book that can only be described by comparing it to the author's cadaver; the Book, if not simply dead, is nevertheless "not living." If the Book lives, it will only be through readings carried out by "dead" people, or people whose "life" is nothing more than sheer repetition.

This hide-and-seek with death continues when Kojève considers what the future existence of "dead" Man will be like. For a basic question remains: who will read the Book? Will it be required reading for citizenship in the universal State? Who in other words are the "men" (lowercase, dead) who must read and understand the Book? Must one read it in order to be dead? Could one slip back into (living) humanity if one fails to read it—if, through one's inactivity, one transforms it back into mere material, binding and ink-stained pages? Would inattentive (incorrect) reading—changing, perverting—mean a fall back into history? Or merely illusion, error?

Kojève never really answers these questions. When he attempts to give some "practical" idea of what posthistorical social life would be like—perhaps thanks to insistent questions from the likes of Bataille and Queneau—he is able to muster two answers. First, posthistorical man runs the risk of simply becoming an animal: if he is merely, once again, "natural," Logos will disappear and human language will not be different in kind from the language of the bees (*ILH,* 436–37). Since this death of Man is clearly unacceptable, Kojève proposes another end: this time the purely formal activities of the Japanese are the model. Noh theater, tea ceremonies, and kamikaze bombings are or were empty (pointless) activities. What is important now is that the labor of negativity continues—a "Subject" continues to be opposed to an "Object"—although "time" and "action" have ceased. Man's labor continues in a void as does the struggle for the recognition of the other. Posthistorical man will be, in other words, a snob.

It seems hard to imagine how these versions of posthistorical language can be separated from a simple death, i.e., retaining the Book only as mere matter, as binding and ink stains: either Man reverts to animal status, or his labor/play is purely formal, like that of the Japanese. But the Japanese were able to evolve this formality without reading the Book: Kojève realizes that they've been snobs for hundreds of years; he recognizes that snobism is blissfully ahistoric.

Thus we have reading as a kind of sheer stuttering, a repetition that saves the Book from being a cadaver only at the expense of putting forth a reader who does not read: man as bee, or as a Kojèvian Japanese who does not even need to read the Book. Man's "essence" is "reducible to this reading," but at the same time the reading is not important enough to figure into any Kojèvian fantasy about how posthistorical Man will live. Man, in other words, is all reading, all the time, the pure Spirit, the unchanging act—and at the same time he is blissfully ignorant of the reading and the Spirit. The Book is both compulsively read, the living act of repeating the same understanding over and over—and it is just as easily sheer matter, paper, binding, and ink. It is absolute Life, Spirit, Consciousness, unchanging—but precisely because it is unchanging it is dead, sheer repetition, mechanical to the point that the Book morphs, with no transition at all, to simple unread dead matter, pulped trees, the cadaver of the Wise Man inevitably forgotten by formalist snobs.

Bataille's take on Kojève's *savoir absolu* is of great importance because it allows him to rethink completely the relationship between this knowledge, or what for Hegel/Kojève amounts to the same thing—a posthistorical divinity—and the Book in which it is propagated, codified, and embodied.

In the "Communication" section of *Inner Experience,* Bataille makes it clear that life is not "situated in a particular point," that instead it consists of "contagions of energy, of movement, of warmth, or the transfer of elements, which constitute on the inside *[intérieurement]* the life of your organic being" (*OC,* 5: 111; *IE,* 94). No longer an "isolated 'being,'" the *ipse* is situated at the interstices of violent movements of energy, vectors of force. "Thus, where you would like to grasp your intemporal substance, you will find only a sliding, only the badly coordinated workings *[jeux]* of your perishable elements" (*OC,* 5: 111; *IE,* 94). This "ungraspable inner streaming" is not only a kind of inner communication, but, as Bataille takes pains to note, a constant opening to what flows out or shoots in.

The *ipse,* the Bataillean self, if we can call it that, is not a high point, not a closed entity embodying a definitive consciousness, but instead a momentary conjunction of coordinated and competing forces, an intersection point, a contingent space of energetic communication. "The glories, the marvels of your life, derive from this bursting forth *[rejaillissement]* of the wave that was tied in you to the immense noise of a cataract of the sky" (*OC,* 5: 112; *IE,* 95).

This violent energy, so reminiscent of the Sadean "transmutations" of nature, is both the origin and end of the human being: the author himself, writing in the first person, is born of the abrupt "communication" of two cells, he is (momentarily) maintained by the violent play of electrical forces, chemical concentrations, and microscopic organisms. His pleasure, and the revelation of the illusory nature of the *ipse,* is a result of the communication with the void, or with another torn-apart being.

When Bataille discusses Hegel much the same movement is at work. If absolute knowledge *(savoir absolu)*[16] forms a perfect circle, as Kojève argued, with its end already engaged at its very beginning, there is nevertheless another element, a supplementary one, we (in a Derridean mode) might say. Against the stasis, the perfectly homogeneous and static post-historical knowledge, the eternal and eternally dead Wise Man embodied in the perpetually self-same Book, Bataille poses a problem, embodied in the reader: himself.

> But this circular thought is dialectical. It leads to the final contradiction (affecting the entire circle): *circular absolute knowledge is definitive non-knowledge [non-savoir].* Supposing that, in fact, I arrive there, I know that I will never know anything more than I know now. (*OC,* 5: 127; *IE,* 108; italics Bataille's)

As we have already seen in Kojève, the *ipse* in Bataille knows all, but it also knows its limits. By knowing that it cannot know anything more, it in effect is able to see the border between what can be known and what must be rejected in order for there to be a finite knowable. "But if, through contagion and by miming it, I carry out in myself the circular movement of Hegel, I define, beyond the limits attained, no longer an unknown but an unknowable" (*OC,* 5: 127; *IE,* 109). Miming the Wise Man, miming God, Bataille the author comes to recognize (impossibly) an element—which he will call in this passage "the night" or "the blind spot"—that is by definition unknowable, that has to remain outside if the inside is to be a coherent and complete space. If the task of knowledge is completed, it can no longer aim at anything other than itself, in the instant. That, however, is the problem: recognizing that, to be complete, it must incorporate everything—because it is a totalizing movement—it can incorporate the definitively outside only by appropriating that which is by definition radically unassimilable, transgressing the limit it affirms. At that (repetitive) instant absolute knowledge is the knowledge of its own impossibility as completed knowledge: non-knowledge.

The problem lies in finitude. Even if Kojève's Hegel would banish finitude (temporality) once the end of history is attained, the end, and the knowledge it embodies, is nevertheless finite because knowledge can't go on, can't continue to be augmented. It is perfect, complete, and, as we've seen, dead—in its incarnation (the Book), and in its reading (always the same, incapable of development). Bataille's Wise Man, the one who really attains Knowledge and Spirit, recognizes that "beyond the limits attained" there is something that opens the possibility of knowledge but that is also definitively resistant to it because it has to be unknowable (circular knowledge must exclude it in order to close itself, be circular, be definitive). And yet there is a paradox: for Knowledge to be truly complete, this unknowable would have to be known, *but as unknowable* (if it were simply known, the limit between unknown and unknowable would be reestablished, with the positing of a new unknowable, another outside). Another version of this unknowable is the question, posed by the miming Wise Man (Bataille's narrator) at and after the end: "Why must there be *what I know*?" (*OC,* 5: 128; *IE,* 109; italics Bataille's). A seemingly mundane question, but one that disturbs the entire Hegelian-Kojèvian edifice: if, at the end, I know all, and know the reasons for all, and still pose this question, I am posing it from the perspective of the unknowable. Why must there be a limit to what I know? We know the Kojèvian answer, we repeat it endlessly (it is the Book)—but the answer demanded is one that questions the absolute limits of the knowable, and that questions why, and how, the unknowable can be definitively separated from the known, when its unknowability is needed for the establishment of the limits of the known in the first place (this is why it is the "blind spot," necessary for vision). A tiny supplementary question, then, one unthinkable for Kojève but that inevitably suggests itself to the prototypical reader of Kojève, the one who "mimes," and in his miming doubles, but with a difference. Or put another way: one who *reads,* rather than simply repeats. One little supplement, one little bit of spacing outside the closed circle of knowledge, just enough to see that the limits themselves depend on what they cannot know and must exclude. "In this way existence closes the circle, but it couldn't do this without including the night from which it proceeds, in order to enter it again" (*OC,* 5: 129; *IE,* 111). A doubled knowledge, one that embraces official knowledge and then goes one step beyond it, knowing the unknown *that remains unknown.* Hence "circular absolute knowledge is definitive non-knowledge."

The circle closes upon itself, always again, but in order to do so it must close upon that which defies and undoes all knowledge. Its limit, its finitude, is now situated within the circle of complete knowledge. The limit that constitutes and that is constituted by absolute knowledge is transgressed by the non-knowledge "under" the edifice, the system, by the blind spot staring out of the void in the structure, and by the violence opening the possibility of the system's dialectical oppositions. It is not that Man is dead and that the Book is alive: the Book itself for Bataille is invested with death, it "communicates" with the void of mortality, all the while maintaining itself as Spirit. Hence the incessant movement of knowledge to non-knowledge in the instant: "circular agitation." Not the stasis of the circle that ends in the Book, in the peculiar Spirit that comes to replace God, but the violence of the conjunction of static, perfect, and perfectly dead knowledge with a "night" that devours that knowledge, that plunges it to its doom, and me along with it. "One last possibility. Let non-knowledge be once again knowledge. I will explore the night! But no, it's the night that explores me" (*OC,* 5: 130; *IE,* 111). My supplementary "I," posing the question that topples knowledge into non-knowledge, is not a stable entity that can somehow be a new Wise Man, write a new Book. At the highest point of knowledge, "I" am explored by the Night, opened out; I enter into communication with the Night, the violent energy of its agitation breaks apart the boundaries that would have constituted a new End point, a new summit. This is a death of Man that doubles Kojève's: it is the death of the Wise Man, and his knowledge. It is the loss of the Kojèvian Book.

> Knowledge is in no way distinct from myself: *I am it,* it is the existence that I am. But this existence is not reducible to it: this reduction would demand that the known be the end of existence, and not existence the end of the known. (*OC,* 5: 129; *IE,* 110; italics Bataille's)

My knowledge is inseparable from, and ends in, existence, this night beyond which knowledge cannot progress, cannot imperialistically appropriate further states of being or historical conjunctures. "The night," like the charged matter that constitutes and ruptures the body, is an end that cannot be put to use, cannot lead anywhere else. Knowledge needs it, but it cannot be reduced to knowledge. It is the internally constituting violence, the energy of the *ipse,* the blind spot that makes vision possible but that swallows vision when it is focused upon (*OC,* 5: 129; *IE,* 110). It is the scattering when communication opens the body to another, or to death; it is the violent Nature of communication, as opposed to the calm, limited, tamed Nature, the standing reserve, always at the disposition of post-Historical Man.

Yet we see Bataille, in his "circular agitation," making conscious the double movement that was implicit, but unthought, in Kojève. There too the Book was caught in a kind of circular movement, not from beginning of History to its end—that was a one-time circle—but in the same moment, from full Spirit to the death of Spirit. Was Man simply an animal, or was he a fully conscious reader of the Book? Was the Book alive or dead, consciousness or mere paper? Was the Book's (absolute, invariable) reading necessary or not? If it was, for whom was it necessary? If not, what then of Spirit? All these questions hover over Kojève's version of the Book, and they are never resolved. From highest knowledge to simple ignorance and back—this is the movement Kojève never explicitly formulates.

Bataille does formulate such a problem, but in doing so he does not simply add on to the Hegelian edifice: he accepts it, affirms it, parodies it. For Bataille, the highest Spirit is also the plunge into "night," the loss or transgression of mastery and knowledge, knowledge as mastery that is loss. The summit, the peak, is also the abyss, swallowing and disgorging. This back and forth movement, which does not take place in (and as) a constructive Time, but rather in an instant of agitation, of uncontrolled and uncontrollable temporality, constitutes the very movement of Bataille's Book, the *Summa Atheologica.*

Rather than dead Spirit, which is perhaps completely alive in its periodic reading, perhaps not—and whose reading is by definition always the same (and in that, we might add, always dead), always correct, never "perverse"—Bataille's Book puts itself forward as the written equivalent of the "intimate world." Bataille's Book can never be nonperverted for the simple reason that it embodies nothing, cannot constitute itself as a simple end point, does not serve as the reason or Spirit of humanity, serves if at all as a hole in which the reader is lost and in which the Book loses itself. Unlike the closed Kojèvian Book, Bataille's Book "communicates" through its wounds, its breaks, with the other, the reader, and with the void. Its alternation between life and death is not unthought, unthinkable; it is instead the principle of this alternation, the uncontrolled shifting of knowledge to non-knowledge and back, a "higher" knowledge that is also a fall into deepest non-knowledge.[17]

A Book then, of movement, of displacement: an intimate Book, not a static repository of truth that makes possible the coherence of Society in and through the completion of History: instead, a Book that "communicates," that opens Sense out to non-Sense, that explicitly alternates between

Knowledge and non-Knowledge, between the social realm in which it acts as a trap and a realm of "night," of continuity and loss. A Book that is not a stable point, a guaranteed and grotesquely unchanging locus of meaning, of awareness, but a Book that moves meaning, and the *ipse* that is inseparable from (its own) meaning, out of itself.

Bataille's Book is not a self-enclosed, satisfied, and satisfying object or Truth: it does not, once and for all, convey the meaning of Spirit, of History, or, for that matter, of a secular mystical experience. It does not set itself the impossible task of conveying the Absolute—Truth or Enlightenment—in the fixed and finite phrases of a doctrine or autobiographical account. It is, instead, the opening-out of this sort of Book, its sacrifice, its communication to the realm of "night." It makes intimate, as object and as Book, the Truth of Philosophy—and of Religion. This is what Bataille is getting at when he states that his (fictional) autobiographical self is "miming" Absolute Knowledge. To mime is to double and to parody, maintaining the model while shifting it, translating and betraying it; the act of shifting also entails the release of an erotic or intellectual blast at the instant Truth is given over to the void, lost in animality. Without stability, the eternal sunshine of the Hegelian-Kojèvian Spirit, there is the discharge of the Bataillean non-knowledge. Like Bataille's formless matter, or like the intimate world, Bataille's Book leads nowhere else, accomplishes nothing. It is the laughter at the end of the intellectual project. And there is no "bad faith" in Bataille simply because there is no possibility of good faith: at the end of (Bataille's) history, one could never choose between *savoir* and *non-savoir,* since they are impossible to disentangle in their "circular agitation." There is no question of erecting a fiction (God) and then getting an ever-frustrated charge out of blaspheming it because it *is* a fiction (Sade). For Bataille, knowledge is both completely necessary and a simulation of itself, because it is dead.[18]

Rather than disavowing the parody of knowledge, the truth of its simulation, one affirms it, revels in it, gets a kick out of the energy released from the expenditure of truth (spent truth, emptied truth, recycled as the always again affirmed simulation of truth, truth with a simulated, parodic, toxic meaning, but [dead] truth nevertheless).

Bataille never lost sight of the Sadean component in this transfer, this sacrifice of sense, even when he was writing in his most Nietzschean mode. For Bataille, however, unlike Sade, this element of "transmutaiton," and of crime, was a way of attaining a "moral summit": blaspheming God nevertheless entailed a miming of God, and as God, or as a parodic God, one could never escape the "summit."

> Men can only "communicate"—live—outside of themselves, and just as they must "communicate," they must want this evil, the stain that, risking themselves, "being in play, makes them mutually penetrable . . ." Now: *all communication has to do with [participe de] suicide and crime . . .* Evil appears in this light, as a *source of life*! It's by ruining in myself, and in others, the integrity of being that I open myself to a commun-ion, that I accede to the moral summit. And the summit is not *under-going*, it is wanting *evil*. (*OC*, 6: 318; ellipses and italics Bataille's)

Sade of course never wrote of a "moral summit"—Bataille takes this orientation from Nietzsche, no doubt, but the Sadean resonance is still clear. As in Sade, the self-satisfied being is broken open to the outside: in Sade it was to the energy of Nature; here it is to a moral summit that is nothing other than the violence of crime itself, the vertiginous downfall of a secure and stable life, the void at the peak. Yet that summit entails a giving to others: a suicidal generosity toward the victim, toward my double, the other. This, in Bataille, is how the other is (non)recognized, or recognizable.

Bataille formulated this paragraph in conjunction with a "discussion on sin," held with a number of Catholic priests, as well as existentialists (including Sartre, Camus, and Simone de Beauvoir), in March 1944. That the priests, and Sartre, were willing to debate him (the discussion was transcribed; see *OC*, 6: 328–59) indicates the extent to which, at this time, both Catholics and existentialists felt challenged by a Bataillean "religion" of evil (and felt obliged to respond to it).[19]

Much was at stake in 1944, and no doubt Father Daniélou, like Sar-tre, imagined that a postwar order might very well embrace a Sadean-Nietzschean doctrine that rejected both conventional Catholicism (many Catholics had, after all, collaborated with the Germans) and conventional existentialism (Sartre's gloomy insistence on lack and sober practice would seem to offer little to a generation just emerging from four years in the claustrophobic "no exit" of boredom and terror).

That Sartre, and not the Catholics or Bataille, won the battle, if not the war, for public opinion need not concern us here.[20] What is important is Bataille's continued emphasis on the notion of God, or Jesus, as a figure of transgression. For Bataille, God is dead—which does not mean that he or she simply does not exist. It is much more complex than that. Bataille himself is not a simple atheist; he does not believe, as Sade purportedly did, that science and reason had somehow rendered God obsolete, or that religion has been subsumed by absolute knowledge and the Kojèvian death of Man. But for Bataille God himself is an atheist: he does not "believe" in himself, least of all as a subservient, limited human construct. In the same way, his Book—the Bible—can be said to "exist" (if at all) only as a dead Book, a Book that does not believe in itself: a counter-Book, Bataille's Book.

God's Book, Bataille's Book

Bataille rewrites the Bible, in a sense, by avoiding it. With all the talk of God, Bataille never explicitly mentions biblical texts. He certainly mentions mystics—Angela of Foligno, John of the Cross—but never a sacred text that is taken by members of organized religion as sacred: absolutely truthful, unchanging, the Word of God, not Man.

Perhaps this is because Bataille cannot imagine God writing such a text. A consistent God, one who takes himself seriously, who is reliable—this for Bataille is not divine, but merely the product of weak humans who demand an exaggerated consistency that they themselves can never hope to obtain.

> God savors himself, says Eckhart. That's possible, but what he savors is, it seems to me, his hatred of himself, to which no other, in this world, can be compared. . . . If God for a minute did not sustain this hate, the world would become logical, intelligible, the fools would explain it (if God did not hate himself, he would be what the depressed fools believe him to be: dejected, stupid, logical). . . . In the fact of saying "all things recognize him as their cause, their principle, and their end" there is this: a man cannot take any more BEING, he asks for mercy, he throws himself, exhausted, into disgrace, like, at the end of one's rope, one goes to bed. (*OC,* 5: 120; *IE,* 102–3)

> God finds repose in nothing, is satisfied by nothing. . . . He does not know the extent of his thirst. And since He *does not know,* He does not know Himself. If He revealed himself to Himself, He would have to recognize Himself as God, but He cannot for a moment grant Himself that. He is only aware of His own nothingness, and that's why He is an atheist, profoundly: he would immediately cease being God

> (instead of His horrible absence there would be only an imbecile, stupefied presence, if He saw himself in that way). (*OC,* 5: 121; *IE,* 103; italics Bataille's)

A Bible written by God, for Bataille, would clearly not be an unchanging, "sacred" text: it would, to the contrary, be the *Summa Atheologica.* If God really were God, not simply the product of humans demanding secure, safe enclosures that protect their Selves, God would hate himself (i.e., his humanly created self) because that self would deny his own Being. For God "is" (like the world) unintelligible, illogical; he cannot believe in himself. What's more, he can never be in repose; like the "circular agitation," and like the atoms of Sade's universe, he spins in constant, self-engendering and self-destroying motion. He is constituted, if that is the word (and it is not), by the gap in his own Being that moves, transfers, and translates any sense, any consistency, in the direction of the void. His "awareness" is the awareness of his own lack of awareness.

God, for Bataille, is the space at the peak of the pyramid, the point at the center of the labyrinth, around which all things are oriented, all projects derive their meaning, the signified of all signifiers. The only problem is that, according to Bataille, if God really is sovereign, unconditioned, then he owes no one anything; he does not stop time or provide rewards or meaning for the benefit of a needy humanity. In and through atheistic expenditure, that central point of divinity, of meaning, is opened out, lanced, depleted; it realizes, it attains its own emptiness, and in doing so it throws back the sacrificial energy that has been invested in it; it sets off in senseless wandering, circular agitation. God, heterogeneous to man's needs and purposes, upends any sacred or social constructs that would maintain his own purported consistency, his qualities, his goodness. If God really is beyond mere determination (permanence, consistency, goodness), the world itself finds its "truth" and "knowledge" not in God's serviceability, but in his death: Bataille's counter-Book, the non-knowledge of Bataille, is emptied as an *ipse* through the "experience" and rewritten as the death of God. Bataille's language is that of the death of God in that his language—*all* language after God's death—can refer to God only in God's profound absence, in the impossibility of reference to God. The death of God is the contamination of language: the Book is the necessity, the knowledge of God, in "his" death. Denis Hollier writes:

> The name of God introduces the equivocal dimension of his presence-absence whose ambiguous play will contaminate all language. For this name which posits the divine as transcendent is the absence of God: *absence of his presence;* but its reverse, or the sacred (as distinct from the divine), is also a mode of the absence of God, this time however in the *sense of the presence of his absence* and of the immanent experience of this absence. (Hollier 1990, 133; italics Hollier's)

My expenditure of strength ("work") can no longer be subordinated to a goal, some larger meaning, with some transcendental signifier—God—in view; it can no longer travel up the scale to serve or lead to or represent God; *now* it is expended for nothing; it is its own end in miming. This is the deleterious "name of God" of which Hollier writes. My sacrifice is no longer directed to God—his space leads nowhere beyond itself, serves nothing. God's death kicks back the force of my desire and effort, sends it tumbling down the side of the pyramid like an Aztec victim; like laughter, which dethrones certainty and divinity, Bataille's literally pointless writing doubles the name of God, incessantly revealing the void at the peak, toppling the summit, expending sense not in a centripetal movement, but in labyrinthine wandering. Bataille accedes to that summit as void, and sense is repetitiously lost in it, like a victim—or a Sadean heroine—plunging into the gaping mouth of a volcano. "Circular absolute knowledge is definitive non-knowledge."

Like the Hegelian-Kojèvian Book, God in his Book runs the risk of being "the cause, the principle, and the end." Such is the version of God encouraged by a reading of Genesis: God creates the world, he is its sole measure, and the truth of its judgment (in sin, after the Fall). But the irony is that, given such a God, "He would immediately cease being God" (*OC,* 5: 121; *IE,* 103) because, if he were cause, principle, and end, he would be a mere projection of the human need for stability and permanence. God would no longer be unconditioned, sovereign, but instead the mere signifier, the desideratum of human weakness. He would serve a purpose. He would be a slave to Man, who in turn can only conceive of himself as serving a purpose (serving God, serving Man). Bataille's God is the laughing rejection of such divinity and of such humanity; he is the rejection of all purpose, of any ultimate signified; his agitation, his incessant death agony, his self-doubt (the *lamma sabachtani* of Christ on the Cross), his absence to himself—all those things are the movement of Bataille's own text, which rejects the stability of any one reading, any one docrtrine or divine-human certainty. All this is not to say, however, that God simply does not exist; a placid, positivistic atheism in that case would simply replace God with a static conception of Man, at the summit, justifying matter and the universe. A God, on the other hand, who does not believe in himself is not the proof of God's nonexistence, but rather the proof that God himself is a mimic, a fool who parodies his own position in order to discredit it, all the while maintaining its empty position. God's atheism is the highest atheism, the proof not of a scientific universe in which all things can be serenely understood, once and for all, but one in which all beings, miming the dead God, can open themselves out in a "communication," a self-sacrifice, that puts in question the very integrity of their own closed being and their well-ordered little universe. A "communication," in other words, in which they as well doubt their own selfish integrity, their own solidity as reasoning animals, as incarnations of posthistorical reason. They are atheists as to their own selfhood, affirming sovereignty at every turn.

The death of God, then, is inseparable from the death of Man. Not the (Kojèvian) death of Man in a higher, always/never read Book, but his death in a counter-Book, the Book of the miming and the overturning of God's knowledge. Human weakness, the demand for meaning and stability, is what creates the eternal, all-powerful, omniscient God and Book in the first place. God's death, "his" own radical atheism, which "he" undergoes *in order to be God* (in order to know all), falls back onto Man. In Bataille God and Man fall together, in and through the recognition of their own finitude, the open wounds of their being. The counter-Book Bataille writes is the script of this movement, the incessant transfer of the mimed meaning of these two majesties, down and out.

Bataille's Book, then, does not attempt to separate out the eternal word of God from the inherently corrupt intervention of Man—a perennial problem for Biblical exegetes, who must decide between the humanly fallible writing of the text, and the divine veracity of the Word.[21] For Bataille, the Book, his Book, is God's Book, but not the Book of the unchanging Word (Bataille "mimes" a self-hating God when he writes his [non]autobiography, the "inner experience" that is neither inner nor an experience of a not-self). It is not Man who corrupts the Book; Man is the one, on the contrary, who would write an unchanging and absolute Book. It is God himself who doubts, who hates, who opens the wound in the Book, the space of non-knowledge, of not-God. Bataille's atheistic Book is therefore closer to God than the Bible that would represent the Word of an eternal and unchanging Divinity. God's doubt is not doubt in a conventional sense that can be remedied by additional questioning, understanding, faith. It is the movement of non-sense, incarnated at the summit by an atheistic God, that passes through the Book, agitates it, prevents it from being a supreme Book, the apathetic and cruel reason of Nature (Sade), the definitive End of History (Kojève-Hegel), or the unchanging, cause and end of Godhood (the Bible). Such is Bataille's autobiographical text, emptied of all the details one would expect in an autobiography, which in its particularity, its finitude, defeats, in the very act of miming, the truth of sensuous and amoral Nature, the finite progress of labor and the definitive elaboration of truth, and the revealed truth of an ahistorical divinity who is the beginning and end of all creation. The static, the eternally meaningful (which is meaningless since it no longer has a job to do), the eternally dead, is opened in all its incoherency to the violence of Sade's violent atoms, to Hegel's infinite boredom after he completed the *Phenomenology* (and realized the nullity of absolute knowledge),[22] to the lack of repose of God in his own self-hatred and self-doubt, to the fragility of the other, torn open with mortal wounds or ecstasy. Bataille, in writing the atheist "mystical experience," moves perfection, stasis, sense, transfers it, translates it, into the realm of loss while maintaining it, doubling its moves. In Hollier's terminology, his language is "contaminated." He transfers sense, loses it, in "communication" with the other, the reader as victim and double. He expends the concepts, the certainties, the truth; his Book, in its writing and incessant reading, is that anguished loss, that challenge, that cry. "I myself cannot be an *ipse* without having cast this cry [to his readers]. By this cry alone, I have the power to destroy in myself the 'I' as they will destroy it in themselves, if they hear me" (*OC*, 5: 136; *IE*, 116).

4 Bataille's City

Elevation, Divine Eroticism, and the Mortal Fall

In the preceding chapters we have seen the extent to which Bataille is able to face and resolve certain problems: how generosity is derived from a model of transmuting energy-charged matter that would seem to dictate radical selfishness; how an ethics of generosity and postsustainability is derived from a model of apparently limitless destructive expenditure; how a religion of the death of God is inseparable from the writing of a "mystical experience" (a counter-Book) that would seem to defy all recording, all writing (the establishment of a religion of the Book). I now turn to another central question in Bataille: the future of the city. In the figure of the city, as Bataille develops it, all of the problems I have examined in the previous three chapters come together. City life for Bataille implies an extreme generosity, intimate expenditure, and the death (the fall) of God situated in the highest moment of modernity: the cult of Man.

Bataille is concerned not only with the fundamentals of economy and with the religious "experience," but with the social inscription in space of both economics and religion. This is evident enough, given Bataille's concern with the economic, religious, and social implications of sacrifice. But a larger question then appears: if there is a Bataillean ethics, and a Bataillean Book, how will they play out, so to speak, in a social situation? There is, following Sade no doubt, something of the utopian/dystopian in Bataille:[1] thinking future society is an imperative for anyone who would rethink economics, religion, and history, as Sade, Kojève, and Bataille did. And very often, utopian thought involves rethinking social space, most notably the city: More, Campanella, Fourier, Bellamy, Le Corbusier, and many others all attempted to reconceive social space, or city space, in the context of their future societies.

For utopians, after all, the task was relatively straightforward: a completely reasonable and healthy society will entail a completely rationalized public or urban space, one that should be eminently describable (if not believable). In Bataille's case, the problem recalls the difficulty he faced when writing his "experience": how can one write about, describe, that which (even more than theologically correct mysticism, if such a thing is even conceivable) defies all language, all description? In the case of social space, the problem is related: how can Bataille envisage a lived space, a city, that has as its "center" a noncenter that is not a cathedral, not a space of juridico-political organization, not a monumental elevation that reassures on a cultural level by apparently signifying permanence, but a locus that is not a locus, a nonrepresentation of that which means the collapse of representation—the death of God? How does one "find" the death of God in the city? How does one "undergo" the "experience" of the death of God as an urban event? And what is left of the modern city after this "reversal" of a peak of confluence, eternity, and reason into a nonlocus of the fall, of the destructive imperative of time? What is the fate of energy in such a city?

Cities of Expenditure

Bataille's "approach," such as it is, no doubt owes much to the Durkheimian tradition of social science in France, one that attaches great emphasis to the revival of social enthusiasm through periodic collective rituals. Durkheim was very concerned with carrying over his anthropological analyses to the analysis of contemporary society: in other words, he wanted to see how modern, anomic society could benefit from the periodic collective experience that presumably would revive the sense of a (secular) sacred in today's Godless world.[2]

Durkheim's theory implies, for the modern world at least, a model of urbanism, for the simple reason that collective festivals demand a public place, a space in which people can periodically gather. Such a space is obviously most effectively and dramatically implemented in cities, where various monuments, fields, parks, stadia, churches, can serve as gathering sites for the revival, or instigation, of public enthusiasm. City festivals and rituals marked the French Revolution, and the French Republican tradition takes very seriously indeed the symbolic role of monuments that serve as rallying sites for Republican fervor. Clearly these urban sites also have a religious basis: one thinks of the Panthéon in Paris, which was the former

Abbey church of St. Geneviève; deconsecrated, it was in a sense reconsecrated as the holy site of the French republic. The deposition of remains in the Panthéon remains, to this day, a solemn celebration in which citizens gather to receive a charge of secular fervor. It was exactly this charge that Durkheim associated with (what he held to be) the fundamentally rational enthusiasm engendered by sacred ritual, even that of the most "primitive" religions.[3]

For Bataille, however, the city presents a formidable problem. From the outset, Bataille poses the civic monument as a kind of ossified and oppressive ideal, a projection of the authoritative and imperious being that humanity has (erroneously) attributed to itself. In a very early article—one of the first published in the review *Documents* (1929)—Bataille devotes an entire "definition" to the word "Architecture." He has this to say:

> In fact, only the ideal being of society, the one that orders and forbids with authority, is expressed in what we can call architectural compositions. Thus the great monuments arise like dikes, opposing the logic of majesty and authority to all the disordered *[troubles]* elements: it was in the form of cathedrals and palaces that the Church or the State addressed, and imposed silence on, the multitudes. (*OC,* 1: 171)

Here we can see Bataille distancing himself from the Durkheimian tradition in the sense that the sacred place—Bataille specifically mentions cathedrals—serves not so much as a concentration point for enthusiasm as it does a point at which discipline is imposed. The "multitudes," no doubt as disordered as base matter, are formed, not only socially but even physically, by the monument. As Bataille goes on to point out in this article, architecture is a drawing-out of the consequences of human evolution, which progressed from the "ape-like" crouching form to an erect human stature. The monument, taller, more rigid, and more vertical than a standing human, is in a sense the ideal of the human, the perfection of the erection of human. Thus, if one "takes it out on" architecture, one is also taking it out on Man (*OC,* 1: 172), for "Man" is nothing more than this apparent difference from the apes ("he" is erect, reasonable, ideal).

This critique implies a back and forth movement between Man and architecture. The latter is the completion of Man, so to speak, the ultimate elevation and making permanent of his stature; architecture boomerangs back, generating Man and constraining him to be erect, perfect, rational. The only possibility of liberation, according to Bataille, lies in the embrace of "monstrous bestiality" (*OC,* 1: 172)—moving backward on the evolutionary path, away from the elevation that is ultimately contrary to human freedom and, one would suppose, human nature. But "Architecture" does not address the question of how, in a modern society full of plenty of bad architecture, this architecture will somehow reverse itself into the void of the bestial. At this point, in other words, the opposition remains a simple one: bad architecture versus good modern painting that, as we learn at the end of the article, serves as the only way out from the "prison warder" *(chiourme)* that is architecture. But, as we'll see shortly, thinking about architecture as an element of the city will provide a way of making the transition from oppressive monument to orgiastic and death-bound city.

Bataille, in another early *Documents* piece, goes on to link the monument to the city. In "L'Amérique disparue" (Lost America), Bataille contrasts pre-Columbian Inca civilization with that of the Aztecs. Bataille writes of the Inca capitol, Cuzco:

> The character of this city was a heavy, massive grandeur. High houses constructed of square, enormous blocks, without exterior windows, without ornament, covered with chalk, gave the streets a sordid and sad appearance.... Cuzco was in fact the center of one of the most administered, most regular States that man has ever created. (*OC,* 1: 152–53)

Thus the monument, the oppressive building, is inseparable from the city in which it finds itself, and the city is inseparable from the oppressive State. This is a model of centralization: all the wealth and energy flows to the capital, where it serves to make possible a State that oppresses its citizens. And at the center of the oppressive city is an oppressive monument. The "temple of the Sun," no doubt also lacking windows, was a kind of prison, the site of secret executions—not the glorious public and spectacular rites of sacrifice that, according to Bataille, characterized the Aztec city.

What Bataille objects to in the Incas, then, is not so much their brutality as their creation of an oppressive *structure.* The latter can manifest itself in architecture, in the city and the State, but what is most important is that it does not oppress by denying people the freedom to be independent agents; rather, their oppression has to do with the imposition of an order that robs them of their violence, their bestiality, their monstrosity. The Aztec city, in fact, restored this violence to them. Bataille's version of Aztec society gives us some idea of the "monstrosity" he advocated in "Architecture"; he cites Prescott, who wrote that in Aztec manuscripts one is "amazed to see the most grotesque caricatures of the human body, monstrous, enormous heads on tiny, shriveled, deformed bodies whose contours are stiff, angular" (*OC,* 1: 155). Moreover, these deformations, we are told, are not due to any artistic incompetence, but instead are a "conventional symbol used to express the idea in the most clear and striking manner."

Aztec architecture for Bataille is virtually an inversion of that of the Incas. While the solid, bunker-like edifices of Cuzco served to hide sacrifice, the Aztec temple raises it up, celebrates it in plain daylight, amid the swarming flies and spattered blood: "The priest kept the man suspended with his stomach in the air, his back bent over a kind of big marker *[borne]* and opened his chest by striking him violently with a blade of brilliant stone" (*OC,* 1: 156–57).

Bataille goes on to note the contrast evident in Mexico City between the "streaming slaughterhouse of men" and the richness of the city, "a veritable Venice with canals and bridges, with decorated temples and above all with very beautiful flower gardens" (*OC,* 1: 157)—flowers that were used, of course, to decorate sacrificial altars.

What is notable here is the passage from architecture to the city. While the article "Architecture" singles out the monument in its oppressiveness, "Lost America," virtually contemporaneous, is at great pains to situate architecture in an urban setting. Cuzco is closed off, oppressive; its citizens are met everywhere by blank walls. Mexico City is open, full of canals, flowers; it too has giant erections, temples dedicated to a vast mythology. But this mythology is not one of a noble, logical, and elevated being; on the contrary, the Aztec gods are hideously and grotesquely deformed.

It would seem, then, that a given architecture cannot be understood outside of its urban context. And the urban is not simply the movement of crowds in streets or over canals; it is the *dégringolade,* the tumbling down the steps, of the sacrificial victim after his heart has been torn out (*OC,* 1: 157). While architecture in itself may be the site of the concealment of sacrifice (Cuzco), the city in conjunction with architecture is the space of spectacular and grotesque sacrifice, the ritual that communicates to the population the monstrosity of their humanity. Architecture in itself is not oppressive; both the Incas and the Aztecs erect giant, solid temples. What varies is the setting and use of the temple in the city.

Thus the city is the space of the oxymoron, the dissemination of the sudden and violent rerouting of the human: the highest human is also the tumbling corpse; the assembly of the multitude before the temple is also the confrontation of the human with the limit that tears it from its totality and throws it into the abyss. The crowd in the city is not so much a union of identities that melds in a larger whole, a larger union; on the contrary, it is the confrontation of each being with the precipice, the frontier over which it hurtles into the void. Bataille's city offers not cultural and political union with social affirmation; on the contrary, it proffers, with the sickly humor of the Aztecs, the higher accomplishment of the human, which is the *conscious* collective (but inevitably individual) fall into the void of death. We should note that the fall is inseparable from an elevation (the height of the temple) and a centrality (the location of the temple in the city).

The Inca versus the Aztec city: this will be a hidden preoccupation of Bataille in the later period of the 1930s. But rather than considering the historical or archeologically derived city, he will pass to the modern city: Paris. He will try to envisage not the city of the grotesque gods, but rather the modern city that would at first seem to be as far from pre-Columbian Mexico City as one could imagine. But the modern city, it turns out, in its very modernity, in its very rejection of the gods and its embrace of hyper-efficiency, swings back to the void around which the Aztecs circulated. In fact it goes beyond them, since it sees the death of gods not as a mere component of bloody ritual, but as the death of God: a fall into radical temporality (the time of loss, of decay, rather than of construction): death and the non-knowledge of expenditure.

First, then, the obelisk. Bataille's analysis of this monument par excellence appears in his article "The Obelisk," published in *Mesures* in 1938. This was a period in which, unlike the later war years and the *Summa atheologica,* Bataille was concerned not only with a "mystical" experience and its larger consequences but with a social celebration of such an experience. At a number of points he fastened on the idea of a celebration of Louis XVI's decapitation, which occurred on January 21, 1793, on what is today the Place de la Concorde.[4] The celebration was to reenact the moment in which the Acéphale, the headless divinity, the mythological figure of the death of mythology, replaced, in Bataille's mind at least, the king as the (dis)organizing myth of French society.

The obelisk for Bataille is as important as the king's beheading. Yet at first glace it would seem that by putting an obelisk at the center of the Place de la Concorde the French Republic commemorated not headlessness, of society or the State, but on the contrary, the very principle of the head.[5] Bataille links the obelisk to the pyramid; in fact the obelisk is nothing more than an elevated, erect pyramid ("the obelisk was to the armed sovereignty of the pharoah what the pyramid was to his dried-out remains" [*OC,* 1: 503–4]). Bataille notes that the pyramid was a replacement for the king's body:

> Each time that death threw down the heavy strong column [the king's body], the world itself was shaken and put in doubt. Nothing less was needed than the giant edifice of the pyramid to reestablish the order of things: the pyramid made the king-god enter into the eternity of heaven next to the solar Râ and, in this way, existence rediscovered its unshakeable plenitude in the person of the one it had *recognized.* (*OC,* 1: 504, italics Bataille's; *VE,* 216)

Here we have a clear replay of the "Architecture" argument, but this time Bataille stresses not only the ideality of the stone erection, but its basic function of *stopping time.* The death of the king is a reminder of time; the pyramid, entombing the king's body, serves as a giant projection of that body, an enormous erectile mass making the body permanent, unshakable. The pyramid not only celebrates the dead king, but it makes him and his power eternal; as Bataille puts it, "history endlessly takes up the response of unchangeable stone to the Heraclitean world of rivers and flame" (*OC,* 1: 505; *VE,* 216). The obelisk carries on this movement, linking the time-stopping, idealizing nature of the pyramid to the omnipotence of the military leader.[6]

Bataille now introduces a movement that had been absent from his earlier essays on architecture and the Incas and Aztecs. Rather than simply contrasting edifice and sacrificial downfall, windowless prison compound and flowered city, he shows how the pyramid/obelisk is inseparable from its own downfall. For the modern city Paris is, in this version, a double city, both an attempt at the establishment of an eternal Cuzco and the city of violent downfall, of the "sensation of time" and the sacrificial movement of the death of God.

The transition takes place because of the nature of *circulation.* In the most difficult section of "The Obelisk," "Mystery and the Public Square"

(*OC*, 1: 502–3; *VE*, 214–15), Bataille contrasts the existence of the isolated life with that of the "agitation of innumerable multitudes." Outside of the "attraction" each human has for another, we are told, they are nothing, "less than shadows, less than dust particles." But their attraction results in an agitation that is, apparently, their truly significant existence: "human life alone," Bataille tells us, which is demanded by the human multitudes, entails "horror, violence, hatred, sobbing, crime, disgust and laughter. Each individual is only a particle of dust gravitating around this acrid existence."

We have gone from a dust particle that is meaningless, that in its own existence is of no interest or power to any other particle, to a mote that is powerful through its attraction and its agitation. No longer alone, it gravitates and moves—or as Bataille would say later, it communicates—with other particles, *around* something. It gives itself by opening itself out. But the something around which it turns is not a monument; on the contrary, it is "nameless" and implies moments of "experience" (for want of a better word) that recall the "monstrosity" Bataille valorized in his earlier essays—elements not that far from the values espoused by Sade (violence, disgust, crime). The obelisk as figure of eternity is reversed and becomes, like the Aztec pyramid, a (non)figure of the fall.

What is most important is that this swarming is *urban*. This is strongly implied at the end of the section when Bataille notes that contemporary swarming is so dense that "many brilliant intellects," not recognizing what these existences are gravitating to, assume that what counts is the individual particle, the individual life. But Bataille states, using a revealing metaphor, these intellects "conceive human existence as badly as he who would judge the reality of a capital according to the appearance of a suburb, who thinks that this life should be considered in its empty and peripheral forms, and not in the monuments and the monumental squares of the center" (*OC*, 1: 503; *VE*, 214–15).

That is quite a leap—from the center as "acrid existence," as axis of crime and disgust, to the center as monument. What is clear from this paragraph is that Bataille would retain the urban concentration of the particle; we get nowhere, he tells us, by concentrating on the individual, as peripheral and empty—closed in and selfish—as suburban life. But then we are forced to ask: how do we make the transition from the center as monument to the center as acrid existence, as "sensation of time"? Such a transition would make clear how modern-day Paris is to be transformed (or has already transformed itself), in effect, from pre-Columbian Cuzco to the Mexico City of the Aztecs.

The transition can take place because the obelisk is an empty symbol. It is both an eternal shield against time and a barrier that means nothing. It is a kind of pure speech act, instantiating power and eternity. But in its very purity, its historical meaninglessness, it becomes fragile, ready to fall. It reveals itself as an empty signifier. As Bataille reminds us in the section titled "The Scaffold" (*OC,* 1: 511–12; *VE,* 220–21), the permanence of the obelisk on the Place de la Concorde was due to the fact that it meant nothing. It recalled neither the king's statue that had been on the Place de la Concorde before the Revolution, nor the scaffold and the king's decapitation, and certainly not other images directly invoking the Republic. It was neutral, a blank image of purity, of the cessation of time. In this sense it is the ultimate icon of modernity, of the meaningless organization celebrating a total and blank ideal.

This void of meaning was both its strength and weakness. Its meaninglessness allowed it to survive many political transitions, but it also indicated that a certain fear of time—of the madness and crime associated with the endless downfall of temporality—had ceased to hold sway over the populace. As the population becomes more modern, powered by fossil fuels, more focused on individualism, on personal security and comfort, the old terror of the void that monarchical or republican symbols kept at bay loses its meaning. The former barrier to the flux of existence becomes a simple marker, a limit, empty in its secularism, its blank humanism. The indifferent traffic of the Place de la Concorde now flows around the obelisk: individuals hurrying on their way to work or play.

But in a sense this total nature of the obelisk serves to reveal once again the violence of time. As in the *Summa Atheologica,* there is a profound link here between a summit and a sacrificial "experience" of the death of God. As we saw in the *Summa,* the death of God is not to be confused with simple atheism, a disbelief in God tied to a cult of the self or science. Here too we proceed through a dismissal of common atheism: the empty secularism of the obelisk, its divorce from all theology and politics, proclaims the absence (not the death) of God through indifference. God is gone because no one cares: the Nietzschean "last man" is everywhere, driving a car, smug, self-centered, a "particle," and the obelisk is just another marker indicating, for navigational purposes, a locale.

It is just then that the reversal occurs. Bataille writes:

> Due to the very fact that they [the secular, republican monuments, such as the obelisk] had become, for the majority of tranquilized existences, more and more useless, empty and fragile shadows, they only remained standing by being ready to fall and, in that way, revealing the desperate fall of lives much more completely than in the fearful obsessions of the past. (*OC,* 1: 506; *VE,* 217)

The monument, and the city, are thus on the point of transformation, and it is the empty secularism of the city that has made this possible. Suddenly the "meaning," the emptiness, of the monuments is reversed; like the "pyramid of Surlej," the rock at which Nietzsche supposedly had the revelation of the "eternal return," which in turn anticipated the "death of God" (*OC,* 1: 510; *VE,* 220), the monument is now "only a marker *[jalon]* indicating the immensity of a catastrophe that can no longer be contained" (*OC,* 1: 507; *VE,* 217). The marker now delineates the space of a modern absolute knowledge; it also indicates, through exclusion, the virulence of the death that now engulfs it. As limit it falls—is transgressed—through the conscious condensation ("communication") of beings in death-bound community. Significantly enough, all it takes is "someone [who is] carried by glory to the meeting of time and its cutting explosion" (*OC,* 1: 506; *VE,* 217) to reveal the emptiness, the downfall, of the monuments, and hence "it is at that moment that death is revealed." Nietzsche revealed the nature of time when he met the pyramid; it follows that it is Bataille himself (as narrator, as mythical figure of the death of God), meeting the obelisk, who reveals the "explosion," the "cataract," of time.

Nietzsche's revelation was in the solitude of the Alps; Bataille's is in the city. After his revelation, his word, the circulating motes—who blithely drive over the very spot where Louis was beheaded—will somehow recognize that decapitation, that fall of time. Suddenly they will coalesce, affirming their own position, and their loss, in the swarm. Their own "communication" with others, in other words, will become a factor; seemingly alone, swirling, they will recognize themselves in the crowd, in the cloud, that turns around a center that is not a center, a central point that is pointless, a moment of the head in which the head falls into the abyss. The central node now reveals itself, is "experienced," for what it "is": immanent, leading nowhere, good for nothing, the impossible point at which all servitude, all purpose, is lost. The highest moment of the "particle," the ahistorical and place-less freedom of the automobile as complete ignorance and refusal of death, flips over into the shuddering of the death of God and the collapse of modernity.

Another model for the modern city, for Paris transformed, is the labyrinth. In "The Labyrinth" (*OC,* 1: 429–41; *VE,* 171–77), an essay written a few years before "The Obelisk," Bataille makes clear the urban nature of the death of God. Laughter is the response of the center to the periphery, of the capital to the outlying boondocks. But inevitably the lack of being represented by hicksville returns against the center; it is not that the center disappears, but that its valence changes, so to speak, and the non-density, the void, represented by the periphery attacks and invests the center at its moment of greatest centeredness. This is periphery not as smug satisfaction (which we saw in "The Obelisk") but as de-centering, as downfall of the organizing, directing center. "Being can be completed, and can attain the menacing grandeur of imperative totality: this accomplishment only projects it with a greater violence into the empty night" (*OC,* 1: 440; *VE,* 176). The accomplishment of being in its greatest, secular plenitude results only in its overturn, in the fall of the center, the monument. The pyramid stays standing, so to speak, but now it is made volatile, its apex merely a charged point[7] around which violent agitation coalesces. Laughter, the movement that robs entities of their plenitude, attacks the center but in so doing sets the structure—the pyramid, the city—in the motion that loses it in the void.

> Above knowable lives, laughter traverses the human pyramid like a network of endless waves that renew themselves in every direction. This reverberating convulsion grasps, from one end to the other, the innumerable being of man—spread out at the summit by the death-agony of God in a black night. (*OC,* 1: 441; *VE,* 177)

The modern city is a pyramidal structure whose "highest" moment is not the comforting stoppage of time or a divine principle giving purpose to all activity. Like the "knowledge" that "mimes" knowledge, the post-Hegelian movement that pushes Hegel so far his knowledge embraces that which it had to exclude in order to constitute itself—here the city, the central and centralizing structure, embraces at the moment of its sheer timeless perfection the very peripheral nonevent, the very emptiness, that it had to exile in order to be a center, in order to erect a permanent monument to itself at its center. Bataille's literally postmodern city is violently convulsed with a laughter that institutes a deleterious time, a death of God and sacrificial communication of beings, at its "center"—a center that can only constitute itself as center by a downfall in which all centers are lost. Bataille's city is nothing less than a condensation in which beings communicate through wounds, a sacrificial center that is a void of death and horror. What is more, it is a center of transition, of passage, of movement, where an atheistic God communicates with erotic force, where the highest knowledge is prostituted in and through animality, "monstrous bestiality."

The model presented in "The Obelisk" is double; the great edifices erected against the horror of the passage of time both seem to fall by themselves, and they fall thanks to a madman whose lantern "projects its absurd light on stone" (*OC,* 1: 512; *VE,* 221). The total secularization of life, its banality due to the elimination of the threat of random death, its transformation into a mere technical construct making possible the mechanized displacement of monadized selves, has changed the meaning of monuments. No longer do they represent an eternity opposing death, since this opposition is no longer seen as crucial. "Human avidity" is no longer oriented around markers that keep time at bay; rather, apparently reversing poles, it now "aspires to what delivers it from established tranquility" (*OC,* I, 506; *VE,* 217). With the decline of monuments, people live the "feeling of time" as a "movement of breathtaking *[irrespirable]* speed" (*OC,* 1: 506; *VE,* 217).

Speed, then, changes meaning, thanks to the revelation of Nietzsche's madman (Bataille as narrator, his myth of "himself"). With the affirmation of the ephemeral, fragile nature of monuments, the speed of the cars that circle around the obelisk is changed to speed as a sensation of time in its aggressive downfall. Just as simple atheism—indifference to a God no longer needed to stop time—becomes the vertigo of the "experience" of the death of God, so too simple speed as movement in traffic, the greatest and emptiest accomplishment of modernity, opens the way to the speed of the composition, decomposition, and recomposition of beings (the decomposition of the cadaver, the mutilation of the madman, the bestiality of the monster).[8] It is the movement from the naive atheism of Sade (nevertheless bearing the heritage, as we have seen, of alchemy)—the materialism that believed that the science of atoms explained all—to the cult of the death of God, proposed by Nietzsche and intensified by Bataille. The solid, placid matter of nineteenth- and early twentieth-century positivism, at the service of, indeed signifying, Man, is subverted by the violence of Heraclitean time. Dust is now volatilized, it swirls so that placid lives are fused in an ecstatic and base "communication," in the decomposition of a furious and unmanageable energy.

Two things strike us in this model: first, modernity is associated with a city of mechanized transport. The particle, the gnat, Man the individual, circling around a vacant center, is nothing less than a motorist. Second, the crucial role in ending this mindless circulation falls to the madman with a lantern, in other words, to the narrator whose analysis "casts light" on the emptiness and consequent fall of the austere and well-ordered civic markers. It is as if we are seeing the whole process from above, schematically: the traffic circulating, the madman declaiming (as in *Zarathustra*), and, presumably, the modern, planned city pitching into the abyss as the particles "realize" their thirst for the "sensation of time." And this sensation of time, this squandering of time, is the energy not of the seemingly efficient propulsion of particles, but of the failure, the limit, of this ideal movement, and thus it is the spun-off and unusable energy, the mortality, of the particles, their opening out to each other in generosity, in "communication," in joy and anguish before death. Their opening to each other, in other words, in the physicality and immorality of a literally postmodern city.

But what will happen "on the ground"? How will the "death of God" be an "event" (an "experience") that stops traffic? How will the dead God, the vertigo of time, mark the space of the empty, infinitely fragile monument? How will Bataille witness this movement, this fall? Earlier, we saw a fall (literally a tumbling down: *dégringolade*) from the summit of the Aztec pyramid. That was the first version of the opening of time, of nonknowledge, in the urban space. But God there, or rather the gods, were only grotesque monsters, boogeymen, almost cartoon characters. Now we must face the death of God in the postmechanized city, and the transition from indifferent "particles" to the agitated witnessing of this incessant death. "The Obelisk" tells us about the death of God and shows us the particles on the move; now, in another text, we will fall into the death of God as one of those torn and "communicating" particles. "The Obelisk" never showed us how one goes from smug particle to *ipse* passing into the abyss of time. We learned the imperative, the postmodern future, but only from afar. Now we will be in it, recognizing and recognized, seeing and falling.

Mme Edwarda: The Death of God in the City

Bataille from the first tells us of the religious implications of his "erotic" novel. We learn from Bataille's "Preface to *Mme Edwarda*" that

> what mysticism could not say (at the moment it said it, it broke down *[défaillait]*), eroticism said: God is nothing if he is not the going beyond of God in all directions *[sens],* in the direction of vulgar being, in that of horror and impurity, and, in the end, in the direction of nothing.... We cannot add to language, with impunity, the word that goes beyond words, God; at the moment we do so, this word, going beyond itself, itself vertiginously destroys its limits. (*OC,* 3: 12)

We know from the outset, then, that the story of Mme Edwarda, written at the same time as "The Torment," an early fragment of *Inner Experience,* will concern itself with God, his death, and his death in language. Mysticism, presumably a mysticism beholden to an official, monotheistic God, is seen to betray God in the sense that a "recognition" of a *dead* God is precisely the point at which mysticism as a coherent practice (labor, planning) fails, falls. Eroticism therefore carries on from mysticism, capable of saying or showing what was only a void in a religious discourse. The word "God" will now be an erotic word, one that—above all other words—challenges the very possibility of coherent language (language as project, representing and acting in the world), rupturing its limits. Hence God, his death, his language-destroying word—in Bataille's counter-Book—are to be found in an urban, erotic setting. In the first edition of *Mme Edwarda,* these words stood as an epigraph:

> Anguish alone is sovereign. The sovereign is no longer a king: he is hidden in the great cities. He surrounds himself with a silence concealing his sadness. He is lurking, waiting for something terrible, and yet his sadness laughs off everything. (*OC,* 3: 494)

The radically unconditioned sovereign is hidden, in the city, but also in the body:[9]

> My two hands gripping the table, I turned toward her. Sitting, she held high a splayed leg: to open better the cleft, she managed to pull the skin with her two hands. In this way the "rags" of Mme Edwarda looked at me, hairy and pink, full of life like a repugnant octopus. I babbled softly:

—Why are you doing that?

—You see, she said, I am GOD. (*OC,* 3: 21)

The witness—the narrator—recognizes the divinity, the spread of God, opened like a Book, as abject (cursed) matter: rags *(guenilles).* But the "rags" also stare back: they recognize the narrator's glance, and this recognition of the other, the double, is in turn mirrored by the "rags'" recognition of themselves as God. But we know already from the "God" section of "Post-Script to Torture"—the supplement to "Torture" (in *Inner Experience*)—that God's recognition of herself is "self"-hatred. God constitutes herself through the disbelief in her totality, her eternity, her majesty, her causation: but what is God if not these things? God's atheism is the fall, but it is also a recognition of a radical non-knowledge and nonbeing: corrosive time, as opposed to eternity.[10] If Mme Edwarda, or her "rags," recognize themselves, they do so through a radical self-hatred, which is nonetheless the highest knowledge. Divine self-hatred is nothing more than the impossibility of the maintenance of what is demanded by humans (and which they themselves can attain only vicariously, through God): love (of man, of God by herself), perfection, eternal consistency, goodness, truth. God constitutes herself only through her doubt, her hatred, of her own totality, hence her knowledge of her own lie, her rupture of the bounds of language. God is sacrilege. She knows, her knowledge is the highest, but it is the knowledge of the gap of her nonbeing, of death, at the core of her divinity. She gives herself selflessly, not to goodness, but to crime (prostitution, public indecency, the profanation of [her own] divinity). This space of hatred, of the lie, is also the space of eroticism: God's genitalia, prostituted genitalia,[11] are a God that looks back at the witness, the secular urban observer, the modern and apparently stable self ("Why are you doing that?") of the narrator. The look is more than recognition; it is a challenge. It is the blind spot, impossibly looking back.

Mme Edwarda, then, is "not herself"—as God. The narrator knows that "death reigns in her" (*OC,* 3: 25). While "The Obelisk" proclaimed the death of God, through the illumination carried out by the madman/narrator, Mme Edwarda presents the scenario of God herself disappearing in and through the public monument. In this case it is the Porte St. Denis, the seventeenth-century archway erected on the Rue St. Denis, the ancient north-south axis of Paris, and the traditional street of prostitution in Paris. It is now God herself who indicates her absence, the emptiness and fall of the monument, at the void of the door *(la porte)* that embodies the duality of God (as in the case of Janus, Dianus, Denis, god of doors, the underworld).[12]

> I trembled at the idea that she could flee, disappear forever. I trembled accepting it, but imagining it, I went crazy; I hurled myself forward, going around the pillar. I turned quickly around the right pillar: she had disappeared, but I couldn't believe it. I rested there, horrified, in front of the *porte,* and I entered into despair when I perceived, on the other side of the boulevard, motionless, the domino lost in shadow: Edwarda was standing, still visibly absent, in front of terrace tables *[une terrasse rangée].* I approached her: she seemed mad, obviously coming from another world. (*OC,* 3: 25)

Through the space of transition, Mme Edwarda disappears; she reappears, "vacant," mad, on the other side, somehow, inexplicably. God is now *in* the monument, she "illuminates" it not by her presence, or by her narration, but by her absence, her word ("God") that silences but maintains all words; the archway is empty, a space of transition, transgression. It is now not the narrator, Nietzsche's madman, who "illuminates" the monument, but God who disappears, dies, in it, under it.

Perhaps most important is the fact that a single witness has replaced the flow of indifferent vehicles, the modern swarm of selves and their motorized condensation: the narrator. We can see now how the monument is emptied and "falls," bringing with it the fall of the human: it is not just the movement of traffic that has emptied the monument, recognizing its emptiness, its sheer place-marker status; it is God who enters that emptiness, disappears, and drags the (formerly) comfortable, naive observer, the narrator, along with her. From a kind of distant overview in "The Obelisk" we have passed into and out of the modern, following God under the emptiness of the arch, and, with her, we fall into madness, death. We can no longer remain indifferent: Mme Edwarda, God, curses the narrator, and with him, us: "I'm suffocating, she screamed, but you, you lousy priest *[peau de curé]*, I SHIT ON YOU *[JE T'EMMERDE]* (*OC,* 3: 26).

The narrator is cursed by God; she is cursing her own priest. He is ejected out of the role of simple detached narrator; she/God shits on him; she attacks him, in other words, with the cursed matter that undermines the eternity of monumental stone. The priest is covered with God's shit, and, in the process, he comes to "recognize" himself as the endless violence and destruction of creation, or at least its acolyte. Now perhaps "Nietzsche's madman" is in a position to shine his light on the monument.

The secular, the indifferent, does not just circle around the dead God; it chases God, it is defied, insulted, dragged into the emptiness of the monument, and shat out again into a region of hideous transformation. But it is also cast in the position of affirming death, maintaining it, as the quotation from Hegel, at the beginning of the author's preface, reminds us: "Death is that which is most terrible, and maintaining the work of death demands the greatest strength" (*OC,* 3: 9).

Perhaps we see some of this force in the third section of *Mme Edwarda.* Mme Edwarda climbs into a taxi and commands the driver to go to Les Halles—the great Paris (meat) market. They don't get far; she climbs out of the taxi, orders the driver into the back seat, and, with the narrator watching, mounts him. As they couple, the narrator turns on the interior light:

> Love in these eyes [Mme Edwarda's] was dead, the cold of dawn came out of them, a transparence where I read death. And everything was tied together in this dream-glance: the naked bodies, the fingers that opened flesh, my anguish and the memory of foam on lips, everything contributed to this blind slide into death. (*OC,* 3: 29)

God's "fall" into death, the overflow of tears that mark her eyes, takes place in a taxi. God has entered the realm of the secular, into a dust particle, a vehicle, and she has opened it out; it too is now a monument to death, stopped in the street. The driver is a "worker," one of those who, according to Bataille in "The Notion of Expenditure," will be liberated by an explosion of base materialism, of expenditure without return. From the earlier position above the "dust" of traffic we have moved into a car, and God, the dead, frenzied God, has moved with us. The car, now stopped and opened out, has become the ultimate signified not of the seemingly stable human particle, but of the death of God. This sacred mortality transfers to the (former) autonomy and mobility of the particle.

Nietzsche's madman turns on the overhead light and illuminates the scene. But it is God who stops the car, commandeers it, turns its energy in the direction of the fall, causes her dead divinity to emanate in an animal coupling. "God, if he 'knew,' would be a pig" (*OC,* 3: 31). This rephrasing of a formula by Breton[13] reminds us that the highest knowledge, God's knowledge, is the void—of hate, of animality, of that which is most foreign to the elevated human. But it is animality with a difference because it is *knowledge.* When the car is stopped, when the driver is mounted (rather than mounts), the energy of the city, of its traffic and mechanized life, is opened out to an overflow, to the continuity of the animal. Body energy now is expended uselessly, not in the circulation of commerce and traffic but in a sacrificial rite; the taxi itself becomes an empty marker, a desecrated altar, thrown open to the copulation and death of God.[14] Nothing could be more archaic than the animal divinity, the all-powerful Goddess, but Mme Edwarda points to the future as well, the time of the collapse of an easy, indifferent modernity—that nevertheless to itself is an absolute knowledge.

The city, like the monument, is now intimate: rather than serving a larger purpose, rather than proposing central spaces designed by traffic planners, it is useless, leads nowhere; it is a concentration point where nothing is concentrated and where we see God, God sees us, and in and through that glance we fall. The mechanized motion of the comfortable self, transported through urban space in order to maximize personal well-being, morphs into the obscene movements of God. Neither appearance nor disappearance, neither concealing nor revealing, the principle of this movement entails an urban displacement that energizes and risks the self, plunges it into anguish *[angoisse],* challenging it and exposing it to time. This is not the homogeneous time of traffic flows, but the deleterious time of eroticism, gender bending, sacrifice, and mortality. The city no longer "serves a higher purpose"; it is instead occupied, traversed by God in the agony of self-laceration and doubt. Space, the emptiness of the sky, opens in God's being, in the public square, in the back seat, in the self passing through monuments split and emptied by divine mortality. Rather than expending energy to transport the elevated self efficiently and painlessly through a streamlined, homogeneous urban space, the lacerated being—the madman, the prostitute—burns off heterogeneous energy, disappearing into the maze of sacrificial nodes.

And here we rest, suspended. Some of Bataille's earliest writings, such as "The Notion of Expenditure," tried to consider the revolutionary, social implications of his theory; as late as *Erotism,* in the 1950s, he was still seeing the revolutionary implications of a "base," subversive (lumpen)proletariat. Certainly, on a different register, *The Accursed Share* attempted to consider the implications, on the level of geopolitics, of a larger "spending without return." But in the areas we have examined in the last two chapters—

the Book and the city of the death of God—it is up to us to imagine the implications.

So the questions remain: What of Bataille's religion (if we can even call it that), his counter-Book, as a force against contemporary religious doctrine? Is there an "accursed religion," as there was an "accursed share"? How to formulate it in an era of Biblical literalism, the era of the Bible as Law?

In a similar vein, the city. For Bataille the city is the locus, the meeting ground of the death of God, the movement of Bataille's counter-Book, with the expenditure of energy without return: the point of sacrifice. Urban space is one of concentration around an elevated marker of eternal divinity—or, on the contrary, it is circulation around, and penetration into, a void in which all divinity is incessantly lost. The Bataillean city entails a movement downward and outward: a fall from elevation, a labyrinthine wandering—or passage through its emptiness and away. It is the endless morphing of the mystical into the erotic, the slippage of the death of God into bodily loss of energy and time. What would such a city amount to, in an era of doomed hypersecularization, when all urban space has become mere marker, mere empty buttress supporting a concrete surface, a space of traffic flow? How is the space of the vehicle to reveal its fundamental emptiness, its fall, at the moment the finitude of the fossil fuel regime is revealed?

The following three chapters will consider these questions directly, attempting to write versions of the future, following Bataille's lead. First, how are expenditure and sustainability to be rethought in an era of apparently limitless waste, an era that has, all along, nevertheless pointed mercilessly to an imminent era of depletion?[15] How will a religion of the counter-Book be lived in an epoch of the apparent collapse of secular narratives of progress (based on models of limitless energy expenditure)? And finally, how can we think about urban life of the (or a) future, given the inevitable decline of easy energy resources and the recognition of expenditure and ritual space in what we have called the "intimate" city?

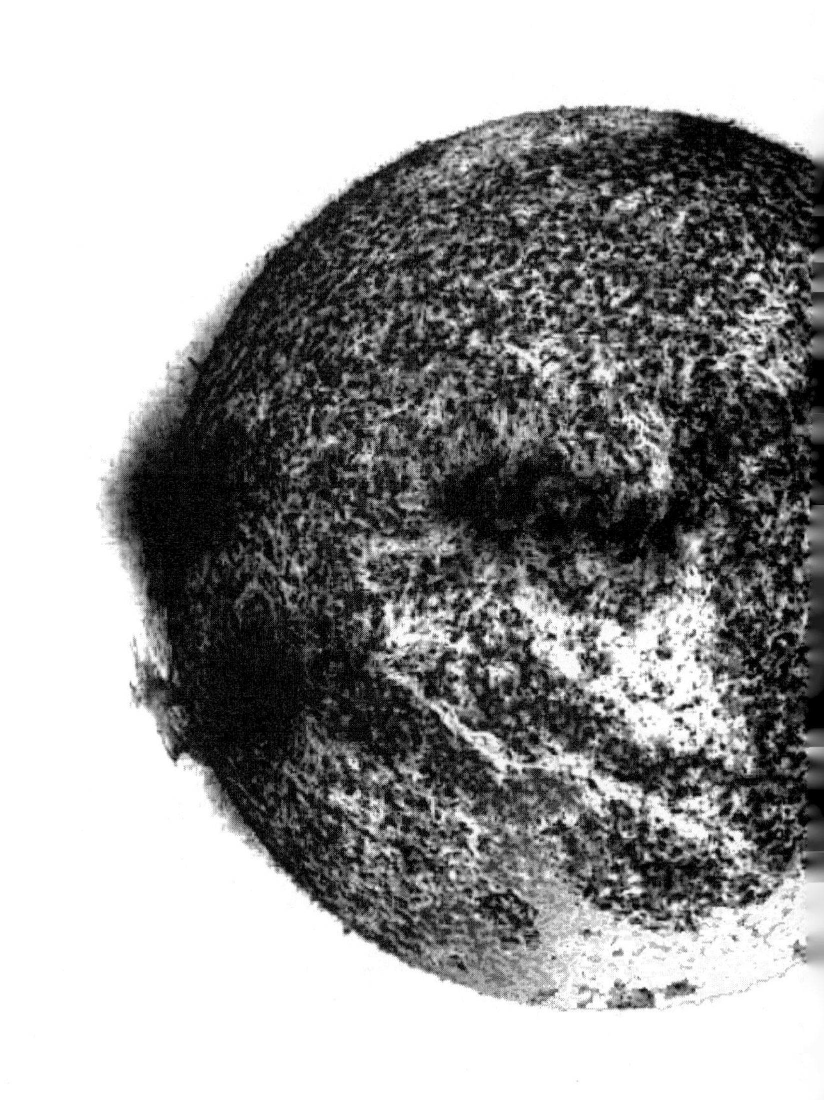

II. Expenditure and Depletion

Orgiastic Recycling

Expenditure and Postsustainability

At this point in the book I would like to shift gears and start to consider not so much Bataille's text, its genesis, implications, and difficulties, but instead the implications of his text for thinking about the future in an era of fast-approaching depletion of the fossil fuel resources on which the "prosperity" of our society depends.

I return to a question posed initially in chapter 2: how are we to think about the "intimate world" of the future as opposed to the cult of Man and the object? Bataille, as we have seen, often posed these oppositions in the context of anthropological or historical givens *(The Accursed Share)*. If, on the other hand, we imagine the future in the framework of resource depletion and ecological crisis, we can, following Bataille, think of tactics of expenditure in relation to models of ecological restraint and parsimonious recycling, as well as to theories of more conventional "wasteful" consumption (car culture). In order to situate Bataille in relation to this future—a future fraught with the promise of crisis, but still seeming to offer some a never-ending mechanized consumer culture—I will examine several works that propose either restraint (simplicity) or consumption as royal roads to freedom. A detour (entailing a significant revision of Bataille) through Heidegger's thinking about technology—already suggested in chapter 2—will facilitate this line of investigation. Finally, I will examine a film by Agnès Varda that, I think, exemplifies the ways in which a Bataillean orgiastic recycling—at variance with both the restrained self of sustainability and the heroic self of consumer extravagance—can be figured.

The reader can be excused if she or he poses the obvious questions: Why all this emphasis on expenditure, on glorious profligacy—isn't this precisely what our society offers in the name of freedom? Isn't the automobile emblematic of the exciting fun we can have, we must have—and that differentiates (or differentiated) us from the Soviet model, concerned as it was with dour economizing and rigorous reinvestment? After all, one can argue that the real "general economy" is that of the modern, industrialized, consumerist world. What can Bataille offer us that is so different? And what is wrong with the car, even if it is a prime version of mass waste? If we have a "need to expend," why not do it by driving over to the mall and buying some stuff?

Is Bataille's expenditure somehow different from what we enjoy in our consumer society? On what grounds could he critique the current culture and its model of energy depletion? On the other hand, if Bataille is somehow "in favor of waste"—isn't that anti-ecological? If he can't really be distinguished from the fun and freedom crowd, how can he have anything to say to ecologists? The cult of glorious waste would seem to be antithetical to the sustainability that many of us deem the necessary foundation of any future economic/social system. If we are entering an era of resource depletion—isn't Bataille pointing in exactly the wrong direction?

Sustainability, Automobility, and the Cult of the Self

Nothing could have seemed more imprudent than the publication of a book in 1949 devoted to the immense wealth of society and the problem of its expenditure. This, after all, was a period of constraint, of deprivation: World War II had ended only four years before, and significant areas of Europe were still devastated. Most of Germany—at least the housing stock and city centers—was rubble, as were many parts of urban England and northern and coastal France. Eastern Europe and the Soviet Union were even more horribly disfigured. Rationing was still in force in many countries, and unemployment—even after the massive death toll of young men in the war—was an acute problem.

Little surprise, then, that *The Accursed Share* sold few copies at the time. Jean-Paul Sartre's approach seemed much more attuned to the situation: this was a period of deprivation, of ruin: the overarching question was how a communist government, or at least a socialist one, would come to power and what role the independent intellectuals of the left would play in it. Essays like *What Is Literature?* and plays like *Les main sales (Dirty Hands)*, pose explicitly both the necessity of a fairer distribution of scarce resources and the brutal role writers and revolutionaries would play in redistributing that meager wealth.

At that time *The Accursed Share,* which argued that the main problem confronting society was plenty, not scarcity, seemed out of place. Other writers around the same time were, however, arguing along the same lines. George Orwell, for example, in *Nineteen Eighty-four,* posed the obvious problem: before World War I, technology had seemed to be on the verge of offering us a new, easier, wealthier life—a more egalitarian one too, since there would be so much wealth it could easily be shared. But two world wars had shown that man was intent on using technology not to increase wealth and make life easier, but to perfect killing machines that would make possible the far greater destruction of wealth. Available resources would be invested in, and labor devoted to, battle cruisers and atom bombs and the wholesale destruction of the world rather than furnishing a car in every garage (hence the endless war in the novel, paired with general poverty). The tragedy of *Nineteen Eighty-four* was that we were so close to reveling in prosperity—nothing really kept us from it—and yet it seemed, more and more, an inaccessible ideal.

As we have seen, Bataille touched on the same theme, but his emphasis was a bit different. He foresaw not Orwell's mournful future in which great wealth, the product of ever more sophisticated technology, was wasted in endless war and repression, but rather a glorious future in which industrial societies would give each other gifts, not least the gift of extravagant consumption. The only problem, as Bataille saw it, was that the other future, not that of drab police state repression but of nuclear annihilation, was also a possibility. Either glorious excess, then, or nothing. A more exhilarating future, but also a more terrifying one, than that foreseen by Orwell.

As it turned out, we got the glorious excess, but not exactly as Bataille imagined it. Bataille, following Kojève, anticipated a new State that would incorporate a Communist-inspired egalitarianism with a freedom to expend associable with the West.[1] A Soviet America, in other words, or an American Soviet Union. But what eventually transpired was quite different—a triumph of waste, but one intimately bound with capitalist and imperialist expansion. Bataille was wrong, but it is important to understand *how* he was wrong. He was not wrong when he saw that the larger, more important question was that of the "uses of wealth" (the title of a proposed book series he was to edit for the Editions de Minuit). He saw quite accurately that the larger problem would not be an endless Sartrian scarcity (always in need of literary intellectuals to drive it further on in a never-ending quest for perfect redistribution), but a profusion of wealth that defied most attempts at understanding. Bataille did not grasp, however, that there are important differences in kind between modes of wealth and modes of destruction (consumption) of wealth. Thus he was unable to see the pitfalls inherent in a theory of expenditure that proposed its highest knowledge—or non-knowledge—in the production/destruction of a modern industrial economy.

This is not to say, of course, that there is not an enormous amount of waste in modern economies. Indeed, they are built on waste; their growth presupposes stupendous, inconceivable amounts of waste and ecological destruction.[2] But a distinction between modes of expenditure (of which waste is only a minor variant) is certainly in order if we are to think its future.

But first: the problem of scarcity. The current moment, the first decade of the new millennium, in which we are attempting to think excess and waste, strangely mirrors that of 1949. Not because we are living amid the ruins of war, but because we face the imminent depletion of fossil fuel resources: what has come to be known as "Hubbert's peak." In short, petroleum geologist M. King Hubbert predicted, in the 1960s, a peak in global oil production for the latter decades of the twentieth century. Hubbert's model entailed applying a bell-curve model, used in predicting the rise and fall in production of oil fields, to world oil production—the projection, in other words, of a model useful for local analysis onto first a national and then a global scale. Hubbert's 1956 prediction of a peak in U.S. oil production was accurate to within one year (1970–71). The chilling lesson of the bell curve is that once production peaks, it declines as rapidly as it increased. There is nothing the "Empire," the globalized world economic system, can do to enable the earth (nature) to produce energy commensurate with its ever-increasing demands. This is the greatest irony: following Hubbert, at least, we can say that energy production will decrease in tandem with a world increase in demand for energy—for a while, at least (Heinberg 2004, 174–79).

This time, in other words, the shortages will not be temporary and easily remedied: we face instead a steady, progressive decline in the availability of resources, a decline that will condition all social life in this century. As Richard Heinberg puts it:

> If [it] were indeed the case—that world petroleum production would soon no longer be able to keep up with demand—it should have been the most important news item of the dawning century, perhaps dwarfing even the atrocities of September 11. Oil was what had made 20th century industrialism possible; it was the crucial material that had given the US its economic and technological edge. . . . If world production of oil could no longer expand, the global economy would be structurally imperiled. The implications were staggering. (2003, 86)

We need not concern ourselves here with why this story has been largely ignored in the mainstream media; this is a matter more for psychologists and sociologists than for social theorists. More important are the implications: that the more or less constant growth in productivity, production, and profits the world experienced over the last century, tracked with a commensurate population increase, based as both were on increases in energy production, is nearing its end.[3] This is the state that Heinberg dubs "Powerdown": "energy famine," a progressive, steep decline in wealth corresponding to the former steep increase, along with the attendant social upheavals implied by this decline.

How to respond? Clearly, lowered expectations are in order, or so it seems. "Sustainability" becomes the mantra of all who would respond to economic decline: sustainability literally means the sustaining of an economy at a certain, appropriate level. This in turn implies not just the conservation of energy resources but their utilization in such a way that they will never be depleted. As much energy will be used as can be produced, indefinitely. Whether sustainability in a literal sense is even comprehensible is another question. Sustainable for how long? At what level of consumption, decided upon by whom? Is a permanently sustainable economy even conceivable? As if sustainability were somehow a Kojèvian end of history, beyond all flux, transposed onto the realm of resource use . . .[4]

Clearly, this dream of sustainability is very different from the current energy regime, which is not sustainable following anyone's definition of sustainability: since all fossil fuel reserves are finite, limited, their use at current levels cannot be sustained. It will eventually end. This means that another energy regime must be established in place of that of the fossil fuels; in order to provide in principle a consistent energy return ad infinitum some sacrifices will clearly have to be made. Solar power, theoretically, can provide energy indefinitely, as can wind power; all other sources are limited.

But, as Heinberg notes, the amount of energy these sources can provide is much more modest than that delivered by the burning of enormous—but very limited—quantities of oil and coal.[5]

Thus sacrifice—not in the glorious sense advocated by Bataille, but in that of "making do, doing without" (the slogan of World War II rationing).[6] "Sustainability" has for this reason become synonymous with a certain morality, one that implies the renunciation of easy pleasure and the embrace of scarcity. A grim determination, blessed by a certain religious tradition, replaces the irresponsibility of the fossil fuel era. Thus Lisa H. Newton's *Ethics and Sustainability: Sustainable Development and the Moral Life* (2003) makes clear the connection between survival in a world of scarcity and the discovery of a vocation. She cites the example of the upwardly striving yuppie whose house and garage are full of professional-grade accoutrements, each of them appropriate for a life never really lived—for show, in other words (88). In the same vein, there is the automobile, purchased not for practical transportation but for living out the multiple fantasies of what one would (sequentially) like to be.

> Half the automobile advertisements you can think of describe not the services of the car or (especially) the SUV, but, as it were, the disservices: how it makes you feel proud; how it towers over the other cars on the block, intimidating the owners; how it makes you feel like you're ramming through the bushes, wrecking the ground cover, destroying mountains; how it, in short, encourages all the anti-social, proud, sinful, and perverse motivations you can access. (82)

Newton counsels simplicity, then, not only as a way of saving the earth—not buying SUVs will help save the environment—but as a kind of vocation that will allow us to discover our "authentic" selfhood (90). Once we have stripped away the blind consumption, the effort to be everything to everybody and nothing to ourselves, we come to discover who we really are: our vocation. "The vocation is the ultimate simplifier of life, because it identifies the one thing that we should be doing and being" (89). Following Aristotle, we acquire the virtues—compassion, courage—by performing them; eventually they will become second nature (87).

All this makes sense—is rational, as Newton would remind us—but one wonders how the irrationality of the marketplace is to be left behind so easily. Simplicity here is just another way of saying renunciation: after we recognize the illusion of the simulacral selves we construct around consumed objects, we can strip down our lives, live "intentionally." Consuming little, devoting our lives to helping others, to being members of the community, not concerned with what others think of our surface appearance, we make possible the dawn of an era of sustainability. As Mr. Rogers would say, we become fancy on the inside, and in so doing consume less, ensuring a world in which the production-consumption-destruction growth racket is finally rendered obsolete.

Yet for all this Newton seems to realize that consumer society offers something that renunciation (simplicity) and authenticity cannot provide. For there is, according to her, if not an ethics of consumerism, at least a kind of aesthetics. She writes:

> Grasping the new thing produces a kind of "high," or rush, exhilaration, some culminating experience for the American of the mall culture, and anticipation of that high is almost better than the climax itself. It is certainly not the "service" of the thing that we seek. (83)

Here is the ultimate immorality of consumerism: there is irrational ecstasy out there, something that verges on the erotic, even foreplay and orgasm (the "anticipation," "the climax"). The "service" of the thing, its rationality, its use value, are shamelessly left in the dust.

An obvious question arises at this point. Maybe there is something profound, something archaic and fundamental in spending. The consumer who constantly refashions him- or herself might be responding to a desire more fundamental than for simplicity or authenticity. Not that it is necessarily authentic—how can it be when the self is lost in a plethora of masks and mirrors? But the demand that sustainability entails a kind of noble renunciation, that it be a castor oil for the soul or that it provide a kind of secular version of the monastic, clearly raises a number of questions. If this is all sustainability can offer, will a significant percentage of the public be willing to embrace it? And if they don't? Does that mean the (self-) condemnation of human society to the hellfire of ecosystem collapse?

Here we could certainly argue that while the waste of contemporary mechanized consumerism *(la consommation)* is *not* the expenditure *(la dépense)* and burn-off *(la consumation)* affirmed by Bataille, there nevertheless is an obvious connection. Bataille himself notes this toward the end of *The Accursed Share* when he argues that there are archaic remainders of expenditure in bourgeois life that have in principle been rigorously extirpated from communism. Here we could argue that while the "intimate world" of Bataille is radically different from a consumerist utopia, nevertheless the latter, in its profligacy, retains a vestige of a more profound expenditure, one that cannot simply be done away with. Waste is, we could argue, a deluded, minor version of expenditure, analogous perhaps to the right-hand sacred as it is opposed, yet tied, to the sacred of the left hand. To argue that the affirmation of sustainability means the renunciation of all sacrificial, atheological, or erotic (in Bataille's sense) urges is to call for a "closed economy" even more rigorous than anything concocted among theorists in Moscow in the 1930s. What this affirmation does mean, however, is recognizing the "tendency to expend" in social life and the difference—and also minor connection—between waste and expenditure.

At least in the Middle Ages the simple life, the accession to authenticity through the renunciation of mirage-like pleasures, was reserved for a noble minority; it was necessarily a minority because most people obviously tended to choose another kind of life. The saint was a saint precisely because her choices were not banal. Now, it seems, we must all don the hair shirt, all renounce the guilty pleasure of waste, in order to save ourselves and the planet. We renounce, we embrace radical austerity, the fantasy of the closed economy—not for the pleasure of another, infinite life but merely in order to prolong slightly, and make a bit less destructive, this one. At least the medieval saints recognized the end of this earth, the final judgment, and a deliverance to a higher, presumably ecstatic, existence. The secular ecological saint renounces only in order to keep on keeping on, and for the earth to do the same.

Even the church fathers, then, the founders of orders whom Newton admires (91), would have recognized the limitation of this vision. Everyone a saint? Not likely. A religion that offers only the thrill of *ressentiment* ("those damned SUV drivers") and the simple pleasure of—let's be honest—the sense of superiority of the authentic self, can hope to have but few converts. People want profligacy, which they identify with freedom, precisely because it is a nevertheless minor, deluded version of a more profound "tendency to expend."

Newton's book is only an extreme form of most writings on sustainability. Virtually all of them preach austerity, warn us that we—the planet—will survive only if we forgo our guilty pleasures, stop wasting so much, and embrace a humble, charitable, "small" life. Charity is certainly a virtue few could argue with, but it seems a bit hard to believe that many will want to renounce so simply and easily the pleasures of consuming and wasting on the grand scale that we associate with life in the "developed" countries.

But we must admit: on one level the advocates of sustainability are absolutely right. We are choking on our own waste, we are destroying the environment just so we can drive those gas guzzlers on six-lane highways to the strip mall. We are thus facing a different, but complementary, version of the "bad expenditure" that Bataille saw threatening the earth in 1949. Any decent work on the current ecological crisis tallies up the numbers: the quantities of unrecycled waste produced, the square miles of land paved over or rendered infertile, the number of species pushed into extinction. The imminent doom. Given all this, even if we assume for a moment the legitimacy of the morality of authenticity that will tempt only a few of us, what does hyperconsumption offer us? What is the freedom it delivers, which we (evidently) find so irresistible? And how can we even hope to resist, or reconfigure, this freedom?

The austerity-authenticity-sustainability school of social commentary is symmetrically matched by another, which defends the very status quo that seems so indefensible. These writers see little problem with lack of sustainability in a social or productive system or (put another way) with a profligate waste of resources; what they celebrate is the very extravagance that, to others, is guilty precisely because it is not accountable. Not surprisingly, this "bad duality" is linked to an affirmation of the freedom and autonomy of the self. This is not to say, however, that this extravagance is not fully, rationally grounded, or that it cannot be defended by having recourse to the history of philosophy or to the history of the American cultural experiment. Extravagance there may be, but it is grounded in a highly problematic notion of subjectivity.

The first case we might mention is that of Loren E. Lomasky, a philosophy professor at Bowling Green State University in Ohio. Lomasky wrote the first, opening shot in the intellectual pro-automobile reaction: "Autonomy and Automobility" (1995). This essay, along with works by authors such as political scientist James Q. Wilson and economist Randal O'Toole[7] (2001) attempt to relegitimize the automobile, after works such as Katie Alvord's *Divorce your Car* (2001) and Jane Holz Kay's *Asphalt Nation* (1998) had attempted to demolish the car's ethical legitimacy. But Alvord and Kay focus on the sheer destructiveness of automobility and the needlessness of most driving. Lomasky, by contrast, argues that the automobile, more than any other transport mode, furnishes us with *autonomy.* His argument is less a pseudo-practical one than a larger philosophical one, focusing, ultimately, on what it means to be human. Following Aristotle (Newton's *maître à penser* as well), Lomasky argues that what is human in us—what distinguishes us from the animals—is our ability to choose rationally: choices thus "flow from and have a feedback effect on our virtues and vices" (1995, 9). To choose, in other words, is to be ethically responsible—again, something animals are incapable of.

Now freedom of choice, our glory and burden as humans, entails a transformation from "a state of potentiality with regard to some quality to the actual realization of that quality"; that transformation is traditionally deemed to be motion. Motion in Aristotle, as Lomasky reminds us, is "ubiquitous because everything has a level of highest possible self-realization toward which it tends to progress" (1995, 9). Movement, then, is human fulfillment, to the extent that free choice is the kind of movement particular to humans. To be human is to be free, and to choose morally (i.e., rationally, responsibly) is to be human in the fullest sense of the word. Lomasky thus invokes Kant as well: as humans we are threatened in our very being by "conditions—manipulation, coercion, intimidation—that impede . . . authorship" (11). Authorship—responsible action—is denied by any force that goes against our ability to choose, to move. Movement is freedom; movement is progression toward the human; movement is the human progressing toward ever greater fulfillment. And, it goes without saying, movement is driving in your car. The car is the device allowing one the most autonomy, that is, the fullest freedom in the "authorship of one's own actions" (Lomasky 1995, 12). To be human is to be a car driver: "Insofar as we enjoy autonomy, we are free beings who thereby possess a worth and dignity that sets us apart from the realm of necessity" (12).

Of course Lomasky begs the most basic question: does the car really deliver the greatest autonomy? Doesn't it offer merely the greatest autonomy in a public realm—the modern freeway-split city—in a milieu that has been (badly) designed for it? Might not the human subject attain a greater level of autonomy in a different urban environment, one that fosters more ecologically friendly—and thus ultimately more satisfying—modes of displacement? Apparently not, for Lomasky at least.

But these questions are secondary; Lomasky does not even consider questions of waste and resource use. Nor does he seriously consider the competing claims of different types of autonomy. Instead, his emphasis is on freedom defined as simple motion, presumably on a freeway; and in this case, the human is most free, most autonomous, in a car. She is most an author of herself: an autobiographical author. For Lomasky, then, the automobile "stands out as the vehicle of self-directedness par excellence" (1995, 24). The self directs itself toward ever greater freedom, responsibility, and thus selfhood. It directs itself to a car.

Strange it is that this autonomy—we could also call it "authenticity"—is essentially the same value celebrated by Newman but in exactly the opposite sense: away from simplicity and in the direction of the seduction of the car and its culture. For Newman real motion was toward the self stripped of pretense, and that meant stripping it of its wheels. The problem, we could argue, in both of these approaches is that they justify car loving or car hating through appeals to the self: its freedom, autonomy, authenticity. But they never question the self. In the case of the car lovers, the self implies motion: but if the self is always (or should always be) moving, to what extent can we even speak of a coherent self? At least Newman's model provided us with a self we could get back to, that presumably always was in residence, even if it was obscured, lost, in the midst of all its faddish whims (SUVs . . .). Lomasky's self is always on the road, always embracing new things, to the extent that we might almost say that its selfhood consists of the motion away from itself, toward something other, in an endless series of simulacra.

This is in fact the problem ultimately posed, if not faced, by another celebrant of modern car culture and suburbia, David Brooks. Brooks, a well-known conservative editorialist who writes for some reason for the liberal *New York Times* (do conservative papers employ liberals?), is as unconcerned with the moral dimension of suburban sprawl as Lomasky is with the ecological implications of his endlessly cruising self. While the latter appeals to Aristotle and Kant, Brooks appeals to the American Dream, which he argues manifests itself nowadays in the suburb: "From the start, Americans were accustomed to thinking in the future tense. They were used to living in a world of dreams, plans, innovations, improvements, and visions of things to come" (Brooks 2004, 255).

The dream translates into constant motion: the continent is, always has been, infinitely rich, and the possibilities are endless. Americans always move because there is always the dream of betterment, self-betterment. America was the opposite of Old Europe, with its stagnation, its classes, its internalized limitations, the repression it foisted on enthusiasm. If Europe was cynical acceptance of mediocrity and stagnation, America was a utopian affirmation of the possibility of change and progression.

> In most cases, people launched on these journeys [across the continent] because they felt in their bones that some set of unbelievable opportunities were out there. They could not tolerate passing out their years without a sense of movement and anticipation, even if their chances were minuscule. (Brooks 2004, 262)

As with Lomasky, here too we see the glories of movement. But changing places for Brooks is not so much the human ideal as it is the American ideal. Perhaps Americans, through their military operations, can pass along to others their love of self-transformation, of freedom as they define it; but it is first and foremost *their* value, what they bring to the world. And that movement, today, Brooks associates with suburbia.

Rather than a stagnant, conformist hell, the alienated breeding ground of teenage shooters and isolated housewives, suburbia is, according to Brooks, the last, greatest embodiment of the American ideal. Americans are always moving, always shoving off for some promised land, and that is what suburbia is: the split-level is only the temporary stopping point before one takes off again, moves to another suburb, another point from which to launch oneself. "This really is a deep and mystical longing" (Brooks 2004, 265).

If movement for Lomasky is rational, and rational fulfillment, for Brooks it is a little bit irrational. The settlers set out not knowing where they are going, and their movement is a kind of blowout, a wasting of tradition, of knowledge, of all that was true but safe and bland. This "mystical longing," this exurban will to power, is not moral in the conventional sense, but wild, reckless, destructive:

> It [what Brooks calls the "Paradise Spell"] is the call making us heedless of the past, disrespectful toward traditions, short on contemplation, wasteful in our use of things around us, impious toward restraints, but consumed by hope, driven ineluctably to improve, fervently optimistic. (269)

So we have come full circle, back to the question of sustainability. But Brooks informs us that our greatness is precisely in wasting: it is a kind of index of our genuine Americanness, "our tendency to work so hard, to consume so feverishly, to move so much" (269). The self is now happy to drive—to find its ethical accomplishment not in freedom and autonomy, but in a feverish squandering we neither can nor should control. And yet there is a problem here, a fundamental contradiction that Brooks never acknowledges. This reckless self that is both the beginning and end of the American quest—that produces, incongruously enough, the placid suburbs and the humble minivan—is also at war with itself.

> What matters most absolutely is the advancing self. The individual is perpetually moving toward wholeness and completion, and ideas are adopted as they suit that mission. Individual betterment is the center around which the whole universe revolves. . . . This is a brutal form of narcissism. The weight of the universe is placed on the shoulders of the individual. Accordingly, in modern American culture, the self becomes semidivinized. . . . It is our duty to create and explore our self, to realize our own inner light. . . . Such a mentality puts incredible pressure on the individual. (276)

The hell-bent wastage that Brooks celebrates—not only of the environment, but of tradition, the family, everything—turns against the self, "puts incredible pressure" on it. It does so for a reason: the self here, in its very movement, its racing forward toward blissful autonomy, is, precisely, never fully autonomous. If it were, it would not have to charge onward. The pressure comes from a profound contradiction making up the self itself: the utopia of the self is movement toward a plenitude of the self, which by definition can never be attained. The American self can only be itself when it is not; or, put another way, the self can only be itself at the cost of not being itself. Brooks's self, his highest ideal, is really the unrecognized empty space of the ever-absent self.

This is not an existentialist dilemma, because this self is less the constructive, project-oriented one of Sartrean labor than it is a destructive, never content, obsessive one: it is engaged in relentless burn-off and can't seem to help itself. And yet, as Brooks informs us, "Everything is provisional and instrumental" (278). Instrumental because everything serves a purpose: the relentless movement of the self in its vain quest to find itself—which, if it were to happen, would result only in the self's loss.

Wastage linked to relentless instrumentality: this is the curse, or blessing, of modern America, of America *tout court,* depending on one's perspective. For Brooks it is a blessing, but one wonders how carefully he has read himself. Autonomist subjectivity is the ultimate signifier of the human here—the American human—and everything is burned, raced through, razed, in the process of elaborating it. But the self itself is only an instrumentality, always leading to something else: the self that it is not. And so on to infinity: everything destroyed to serve a purpose, but the purpose is not just inherently remote; it is by definition unattainable.

Thus Brooks defends car culture, suburban culture, sprawl, the destruction of resources on a scale hitherto unimaginable in human history. Using a cold-eyed profit-loss calculus, we could say: not much return for the investment. A world of resources pumped and dumped for the pleasure of the unattainable self.

But it is, as Brooks makes clear, an invigorating chase, or at least an entertaining one, and this alone would seem to justify it. Freedom, movement, always again blasting off, the exaltation of Dean Moriarty in *On the Road.* Dean makes strange bedfellows with the neocon readers, the developers and highway planners Brooks is seeking to bless. Or maybe not.

We have moved, then, from authentic self to free self, to brutal, narcissistic (autonomous, automobilist) self, to deluded, unattainable self. Always a self: it seems as though the ultimate player in the saving or squandering of human resources is the self, whoever or whatever it is. To save it, to nurture it, to let it bloom in its full humanity, its Americanness, or its authenticity, we make use of resources, nature, either by saving it or spending it. Resources are the currency by which the self is either maintained, elaborated, or set in motion, in freedom. Saved, used, or wasted, resources are the means by which the true human is uncovered, recovered, or discovered. In simplicity, or in driving to the burbs. Man is dead? Not if there are still fossil fuel resources to conserve—or burn.

This leads to a larger question: is there something in the drama of sustainability versus the suburbs other than the health of the self or its drama? If so, how can we formulate it?

Heidegger, Bataille, and Modes of Expenditure

The German philosopher Martin Heidegger (1889–1976) was no doubt one of the European thinkers of the twentieth century, along with Bataille, most concerned with the implications for society of development and energy use. This might come to some as a bit of a surprise, because Heidegger's writings on technology are rarely (if ever) considered from the point of view of energy use and conservation. But in two critical essays of the 1950s, "The Question concerning Technology" and "The Age of the World Picture," Heidegger explicitly formulates the relation between man, the revealing that man effects and that effects man, and the energy that is either revealed or wrested and stockpiled through technology (itself a version of revealing).

It is crucial to understand that for Heidegger there are essentially two modes of revealing, one associated with ritual, pre-high-tech modes of civilization and the other with an intensive "commanding" that, according to the military metaphor, seems violently to force resources from the earth, make them stand at attention, and then stockpiles them, putting them in a barracks, so to speak, before sending them into the battle of consumption.

For Heidegger, *physis* and *poeisis* are intimately linked. *Physis* is the movement of *poeisis,* bringing-forth, by which a living thing "presences" by means of a "bursting open belonging to bringing-forth, e.g., the bursting of a blossom into bloom, into itself *(en heautoi)*" (Heidegger 1977, 10). Another variant of bringing-forth is that carried out by the artisan who makes the ritual object, such as a silver chalice. In this case the ritual object has "the bursting open belonging to bringing-forth not in itself, but in another *(en alloi),* in the craftsman or artist" (11). There are four ways of causing, of being responsible for the thing (6); the craftsman brings them together, and the result is the bringing-forth of the chalice. These four elements consist of the material cause (the material, the silver metal); the formal cause (the form of the chalice); the final cause (the end, the ritual in which the chalice is to be used); and the effective cause, of which the chalice is an effect (the craftsman). This fourfold model of causation sees the role of the human as only partial: there are four types of revealing, and human activity is only one of them.

It is worth noting that for Heidegger energy plays a prominent role in this revealing, despite what this fourfold model would seem to indicate (energy expended in the making of the chalice is not included, interestingly, as a cause). Heidegger later cites the case of the windmill, which "does not unlock energy from the air currents in order to store it" (14). In other words, the windmill, no doubt consonant with the tools of the craftsman, serves only to "bring forth" the energy needed to carry out an operation—the milling of wheat—which is itself a bringing-forth of the wheat as flour. In the *poeisis* of the craftsman or miller, then, energy does have a role, but it is an immediate instance of energy, not stockpiled or accumulated. It is directly applied. In the case of technology *(techné),* however, there is a movement in which

> Man, investigating, observing, ensnares nature as an area of his own conceiving[;] he has already been claimed by a way of revealing that challenges him to approach nature as an object of research, until even the object disappears into the objectlessness of standing-reserve. (19)

This ensnaring is not so much a product of man's will as the aftereffect of a *certain kind* of revealing. Likewise, man himself as subject is an effect of this mode of revealing. In the case of the craftsman making the chalice, there is an awareness that the human is only one factor in the bringing-forth of the (sacred) object; now, in the realm of *techné,* man takes himself to be the author of the process, a process carried out exclusively by him, on nature, for his own benefit. Heidegger mentions "The Rhine" as title, and subject, of an artwork—a product of *poeisis* as sacred bringing forth—in the poem of the same name by Hölderlin. He contrasts this with "The Rhine" as the source of hydroelectric power or as an "object on call for inspection by a tour group ordered there by the vacation industry" (16).

Techné is a form of revealing, and as revealing it is a bringing-forth. This is the movement of the craftsman. Modern technology is also a revealing, but not a bringing-forth. Thus technology is inseparable from *techné* in its original sense, but it is also quite different in that it is a kind of confrontation, a military ordering, a regimentation.

> The revealing that holds sway throughout modern technology does not enfold into a bringing-forth in the sense of *poéisis.* The revealing that rules in modern technology is a challenging *[Herausfordern],* which puts to nature the unreasonable demand that it supply energy that can be extracted and stored as such. (14)

Ensnaring entails a "monstrous" entrapment of natural energy. The Rhine becomes a standing reserve of energy. *Techné* entails not a bringing-forth but a "challenging-forth." Heidegger writes:

> That challenging happens in that the energy concealed in nature is unlocked, what is unlocked is transformed, what is transformed is stored up, what is stored up is, in turn, distributed, and what is distributed is switched about ever anew. (16)

Nature has lost what we might call its autonomy; its model is no longer the bringing-forth of the flower bud, or the energy of the windmill (which "does not unlock energy from the wind currents in order to store it" [14]), but the violent, commandeering, ordering, and stockpiling of energy by the human as challenging-forth. The human, now revealed as a sort of martial monster, is opposed, in its actions, to the bringing-forth that best characterized *poeisis* (a causal model in which the human plays only a part).

And, Heidegger makes clear in another essay, "The Age of the World Picture," reality itself in and through technology can only be grasped as a standing reserve, ripe for quantification, stockpiling, use, and disposal, if it is isolated in an objective "picture," a coherent, passive, inert totality whose only aspect is that it can be brought-forth, by man, violently, in *techné*. "To represent" objectively (as the Rhine is represented by those who would harness its energy) is "to set out before oneself and to set forth in relation to oneself" (Heidesser 1977, 132). "That the world becomes picture is one and the same even with the event of man's becoming *suiectum* in the midst of that which is" (132).

The rise of subjectivity, of the isolated, active self, conquering nature, storing its energy, is inseparable from the appearance of an "anthropology" through which "observation and teaching about the world change into a doctrine of man" (133). Or, we might say, observation and teaching about the world become observation and teaching about man: the measurement of nature's resources and their stockpiling—and wanton expenditure—are inseparable from the stockpiling and wastage of the human in techno-scientific methods. Man the subject for whom the objective world exists as a resource is quickly reversed and becomes man the object who, under the right conditions, is examined, marshaled, and then releases a specific amount of energy before he himself is definitively depleted. Although Heidegger does not stress this point in "The Age of the World Picture," he does make this point elsewhere, noting what for him is the inevitable link between the transformation of the world into a giant energy reserve and the transformation of man into a resource to be exploited in, for example, concentration camps.[8]

Subject/object; this is the infernal duo that, for Heidegger, characterizes modernity. The world is quantified in order to be exploited by "man," but man himself is a consequence of this mode of expenditure. The man who hoards, who works to preserve his individual existence and protect it from all threats, is inseparable from a natural world completely transformed and rendered "monstrous" by a kind of instrumental mania. Man himself becomes a resource to be scientifically investigated, fully known, perfected, made fully human (with an identity and consciousness) and put to use.[9]

This brief excursion through Heidegger on technology is useful, I think, to put the work of ideologists of suburbia and car culture, like Lomasky and Brooks, in perspective. We could argue, following Heidegger, that their version of car culture inevitably entails a subjectivity, one that, as in Heidegger, is both produced by their model and in turn produces it. The illusion "Man" derives his "freedom" from the quantification and commodification of natural resources: oil, to be sure, but also the steel, plastics, and other materials that go to make up the "autonomist" lifestyle. Utility as the autonomists conceive it is inseparable from a freedom that *wastes,* though they are notably reticent when it comes to discussing the consumption of resources on which their favorite lifestyle depends. Heidegger, although he does not explicitly pose the question of waste, certainly implies it: the Rhine, ruined by all those who exploit it, is a "resource" that has been squandered for the self-satisfied pleasures of domestic life and tourism.

I have discussed the analyses of Lomasky and Brooks at such length because they are the most articulate and coherent defenders of the current culture in which we (attempt to) live. These proponents of the ideology of the current American fossil fuel regime valorize a lavish and ruinous wastage but do so in a way that masks it, invoking as they do utility: the squandering of vast amounts of wealth is necessary, indeed is a given, because we are necessarily engaged in developing to the fullest our nature as autonomous, free, individuals. As those free individuals we are the highest being on the earth (as Aristotle would remind us), the most developed. And as such we can be expected to reject any calls to conservation or sustainability. Heidegger, however, would note that our being, our subjectivity, is a quantifiable term that is a function of the very same movement, the very same bringing forth as *techné,* that renders the world a quantifiable mass ripe for exploitation. And such a subject, immediately transformed into an object, a standing reserve, warehoused in an institution (concentration camp, prison, army, hospital, school, freeway, suburb), is itself ripe for use and disposal.

The vaunted subject of the autonomists is for that reason autonomous only in its slavery to a "monstrous" energy regime. Energy is surely wasted in a challenging, but it is a wastage that goes hand in hand with the production and wastage of a subjectivity that is closed in on itself, concerned with its own comfort, stability, and permanence. The freedom of car culture, of the fossil fuel era, is the freedom of a subject whose imperial grasp is inseparable from its weakness as a quantifiable "dust mote" (as Bataille would put it).

Once we have seen the fundamental cult of subjectivity on the part of the autonomists, we can return and consider the model of subjectivity of the sustainability partisans. For them too the chief raison d'être of their model of the future is a subjectivity. Now, however, subjectivity entails not so much the lavish expenditure of a stockpiled energy (cars, freeways, consumer waste) as it does an even more rigorously stockpiled resource base. While Heidgegger's retro-grouch analysis implied a wanton destruction of the stockpiled energy base (the concentration camp as extreme and no doubt self-exculpating example), the sustainability proponents imagine a standing reserve that would somehow not deplete but rather conserve the resources that go into it. "Humanity" would appropriate and store those resources in such a way that they would be perpetually ready to hand. But nature would still consist of a reserve to be tapped and resources to be expended; the goal of the operation would still be the furthering of the stable human subject, the master of its domain. Now the world is really to be useful, and nature is to be pristine exactly to the extent that that untouched state furthers man's permanence and comfort on Earth. The quantified, mechanized destruction of Earth becomes the quantified, mechanized preservation of Earth.

No doubt the sustainable future as sketched out by moral critics such as Newman would be preferable to the dystopian future that would result from a continent completely chewed up and covered with sprawl as celebrated by Lomasky and Brooks. But the sustainable vision is to the autonomist vision what Calvinism is to High-Church Anglicanism. The autonomous, overweening self is consecrated in its subjectivity not through a wild ride on the freeway—which might give the semblance of extravagance and freedom—but through the virtuous sense of renunciation one gets from darning one's socks or writing on the backs of envelopes. The world is small, small is beautiful, it is a prosperous way down, and we will be content, we will be superior in our lowered expectations. We will save the earth from the destruction mandated by the profligate autonomists only by a frugal renunciation that will be sober, clear-headed, modest. The wildness, the irrationality, the aggressive ecstasy of James Dean in his Porsche—heedless, death-driven, glorious—will be, thankfully, discarded.

This makes clear, I think, a weakness in Heidegger's argument. While his analysis is eminently valid when he notes the ties between fossil fuel consumption, stockpiling, and the production and destruction of "human" subjectivity, he loses sight of the difference between the stockpiling effort that would gloriously destroy the (momentarily) preserved wealth in order to achieve the subjectivity effect, and the stockpiling that would spend in only the most restrictive, self-denying way: the least glorious, in other words. Both favor a subject/object model that takes for granted a stable and overweening subjectivity, but one that at least allows for a vestigial exercise of the old, exuberant "tendency to expend." This is the very same difference noted by Bataille in *The Accursed Share* between the capitalism of the classic Protestant ethic types (Ben Franklin, the Calvinists, etc., examined in section 4 of his book)—those who saw, like the sustainability proponents, salvation in their economizing, their constant reinvestment—and the more modern capitalists, whose relatively exuberant expenditure stood out against the dreary and phantasmic reinvestment ethic of the Communists.

The strength of Heidegger's argument is that it shows us the connection between extravagant expenditure in a mechanized, technological economy and restrained, parsimonious spending in a sustainable one. Both economies are fundamentally technological, involving the standing reserve and the basic role of the subject-object opposition, in which the integrity of the subject, the self, is guaranteed through the mechanization of nature and the preservation and quantification of energy resources. Both are dependent exclusively on a conception of energy as a "power to do work," what we might call a "homogeneous" energy whose very identity is inseparable from (apparently) useful labor.

Bataille in effect makes possible the revision of Heidegger in one very important way. Heidegger's silver chalice is seemingly unconnected with the stockpiling of ore and the use of concentrated energy in fuel. (Ritual for Heidegger appears to be somehow radically distinct from all the material processes of energy expenditure.) Bataille, on the other hand, understands that conservation and expenditure are an inherent part of "ritual" production as well as production associated with the cult of Man and the mechanized standing reserve; what for Heidegger was a bringing-forth that seemed only in a minor way to involve pristine, unstockpiled energy (wind, in fact), in Bataille becomes an expenditure that counters the selfish movement of acquisition. This does not mean, of course, that Bataille is somehow against the reuse of materials, any more than Heidegger is against the use and reuse of silver in the production of the chalice. But Bataille is interested in the *economy* of the excessive part, the ritual or sacred part, nonrecuperable matter or energy, in and through which the self is opened out in "communication." There is, in other words, an economy of the excessive part for Bataille; ritual is always already a matter of the concentration of energy in an object and the expenditure of that energy. The chalice for Bataille is an object that defeats utility and its own object-hood in and through its (mis)use in an intimate moment (sacrificial ritual). Both the fabrication of the chalice (as well as the carrying out of the ritual with which it is associated) and the quantification and storage of energy from the Rhine would be, for Bataille, instances of the use and expenditure of energy. The chalice provides an instance of a lavish expenditure of energy (mining silver takes a lot of work; the chalice is uselessly decorated and finished), as does the use of Rhine energy (electricity used to power a wasteful consumer society lifestyle). Heidegger ultimately loses sight of this connection, and difference, by largely ignoring the relation between energy expenditure and ritual.

But seeing the connection between chalice and Rhine power (both entailing energy conservation and expenditure) also helps us see the difference, one that can be derived from Heidegger and that brings a useful correction to Bataille. Ritual—sacrifice—entails a production and consumption of energy that is not stockpiled or quantified in the same way as are raw materials or energy resources used in industrial society. This energy is not and cannot be simply quantified, measured, and doled out in a Marshall Plan; like the "formless" matter it animates, it does not go to the production of a coherent and meaningful (ideal) universe, be it a universe of God or science. We might call this energy "heterogeneous," in opposition to the energy that is merely the power to do work and generate (apparent) order. This "other" energy is energy of the body, of useless body motion in deleterious time; it is inseparable from the putting into question of the coherency of the body, of the self, and of God, that supreme self. It is energy as the flow of generosity, of the revelation of the void at the peak. It is the energy of celestial bodies, matter beyond or below appropriation by the human.

The energy of the Rhine, on the other hand, as discussed by Heidegger, is quantifiable, and hence can be harvested by a scientific-technical grasping of nature. This latter energy involves an objectification of nature but also an objectification of subjectivity itself (stockpiled subjects as just another standing reserve). This energy is "useful," it "serves a purpose," it enables us to be free by strengthening our autonomous (autonomist) subjectivity. Our self, selfhood, selfishness.

Ultimately, the sacred (or cursed) share of energy is not quantifiable because the "inner experience" tied to it does not entail representation; indeed, as we have seen, it entails the expenditure of a language (in Bataille's counter-Book) that would simply represent a stable (phantasmically eternal) world. Thus "communication" of the self, its opening out to death or to the other, is doubled by the monstrous movements of the body and the disgusting dualism of matter to which the body in turn reacts in and as communication (vomiting, sexual arousal, horror, etc.).

Having said all this, however, we should note that the two energies can never be rigorously separated either. Just as absolute knowledge is and loses itself in non-knowledge (the *limit* of knowledge affirmed and transgressed), so too homogeneous and heterogeneous energy are inseparable: energy is dual, not a singular concept, and in its duality it both founds and overturns, both renders possible and conceivable, and destroys in a void. Bataille's point, one that he himself perhaps loses sight of, is that energy, like knowledge, is both unitary and double and that energy that merely founds and sustains in a coherent, quantifiable fashion—like the energy implied in the accomplishment of the Marshall Plan—is only one version of energy, and it is not the sacred share. It is limited, depletable, transgressed in and through *another* energy. Heterogeneous energy, like cursed matter, can never be depended on to guarantee an autonomous and free self.

The consequences of the necessary Bataillean revision of Heidegger (or the Heideggerian revision of Bataille) are extremely important, and in my opinion were never fully recognized by Bataille himself. If the economy of stable and closed subjectivity is tied to quantification and mechanization—"anthropology" in Heidegger's terminology—then the economy of the "communicating" self does not entail the products, or the quantified excess, of a modern economy. It certainly entails energy, but the fate of energy is very different. What is expended is cursed matter, heterogeneous, charged "filth" and not the useful/fun products marketed in an autonomist, subject-centered postindustrial paradise. Thus Bataille himself was off the mark when he proposed the Marshall Plan as an example of twentieth-century potlatch: the problem was not so much that the Americans were "giving" out of self-interest—ultimately the self always reappears as a limit, as an interdiction, to the continuity of blind communication—but that the gift-giving itself was inseparable from the maintenance of an energy regime based on stockpiling and quantification (a fossil fuel energy regime, in short). Americans were giving away money and finished products, not, say, objects carrying a powerful ritual or sacrificial charge, the "power of points" resulting from the exuberance of muscle power, the anguished "experience" of time, and ecstatic participation in frenetic and death-bound activity.

Heidegger, then, did not recognize, fully at least, the energy-based aspect of ritual. He saw that energy was involved—the windmill—but he did not recognize that the bringing-forth, the revealing, was itself an expenditure of energy. He saw energy expenditure as a central focus of *poeisis* only in the context of *techné* and the militarily ordered standing reserve. Bataille sees fully the economic aspect of ritual expenditure, of sacrifice, but he does not see the fundamental connection, noted by Heidegger, between a "technological" expenditure and a stockpiled, protected, and projected *self*. Had Bataille recognized this connection, he would have been able to distinguish the exuberant wastage of autonomist culture—tied ultimately to the growth economy, consumer culture, and the overweening self—from another expenditure, of the body, of heterogeneous matter, of death as internal, and internally recognized and transgressed limit and end. He would have distinguished the faux constructive spending by and for Man from the sacrificial expenditure of the death of God and Man.

This confusion is fundamental because a number of commentators have come to see Bataille's work as an ultimate celebration of the exuberance of postmodern capitalism. Thus Jean-Joseph Goux, in a brilliant article, argues that George Gilder, a Reagan-era apologist of capitalism, essentially co-opts Bataille's argument; if the bases of capitalism as it is practiced in the West today is risk taking, squandering resources blindly in the hope of a far from assured profit, capitalism will have more in common with the risky potlatch ritual than it has with the miserly savings plans of Ben Franklin.

> One can now point to an "antibourgeois" defense of capitalism, an apposition of terms which resonate disturbingly, like an enigmatic oxymoron. Everything happens as if the traditional values of the bourgeois ethos (sobriety, calculation, foresight, etc.) were no longer those values which corresponded to the demands of contemporary capitalism. And it is in this way that Gilder's legitimation . . . can echo so surprisingly Bataille's critiques of the cramped, profane, narrowly utilitarian and calculating bourgeois mentality. (Goux 1990a, 217)

Thus for Goux, Bataille becomes a Reaganite *avant la lettre,* and the accursed share is not much more than the motivation of every contemporary billionaire.

If, however, we shift our focus slightly, from bourgeois versus "primitive" economy to the difference between the economies of energy regimes, a move authorized by Bataille himself in his emphasis on energy as the fundamental factor in wealth, then we will see that there is a profound difference between expenditure as a feature of the standing reserve and expenditure as it appears as a function of intimate ritual. In the first case, expenditure is tied to the production and maintenance of the self (Brooks and Lomasky would certainly agree); in the second, to the fundamental "communication" of the self in loss, dread, eroticism, death: the intimacy that accomplishes nothing, goes nowhere, but that is inseparable from an "inner experience" (which is neither inner nor an experience). Even if there are (obvious) elements of sexiness or risk taking in mechanized, quantified expenditure (as both Brooks and Lomasky would argue as well), the latter is fundamentally tied not to dread and *non-savoir,* but to the faux permanence and dominion of human subjectivity. Heidegger's critique, which is perfectly consonant with Bataille's, is not so much antibourgeois as it is one that is established against a certain way of conceiving the production, storage, and waste of energy resources. The energy stored in and released from a strip-mined mound of coal is qualitatively different from, for example, the bodily energy discharged at the contact of an eroticized object.[10] Heterogeneous energy is what is left over, in excess, after the other energy has depleted itself, either literally or logically, in the completion of its job. It is there after homogeneous energy is quantified and used to the point of its own extinction, or after it has revealed itself as nonsustainable in the sense that its excess is inseparable from the production and maintenance of an illusory presence (its end is the production and sustenance of a modern subjectivity that is riven, death-bound, but that takes itself to be total, essential).

Thus Bataille's affirmation of expenditure and loss cannot be simply identified with the waste of consumer culture and modern capitalist economies. To be sure, "modern" economies are based on an ever more frantic rhythm of production-consumption-destruction, and in that they are deeply implicated in a wastage process more fundamental than the world has ever seen. The amounts of food, metals, fuels consumed, and the amount of all types of nonrecyclable waste produced are staggering. Whole forests and ecosystems are destroyed without sentimentality in the name of utility (furthering the necessary comfort of Man). But, simply put, this is not the kind of consumption, or expenditure, that Bataille is talking about. At best we could say that they are "bad" versions of expenditure: without any awareness of it, people "waste" because this society has turned its back on expenditure. It is their only option, their only way of spending—and for this reason they would hardly refuse this waste if their only other course of action was a radical conservation from which all expenditure, waste or burn-off, *consommation* or *consumation,* was eliminated.

Our detour through Heidegger indicates that modern subjectivity—subjectivity that itself can be objectified, quantified—is inseparable from an instrumental conception of "resources": matter is now quantity—measured, hoarded, and then spent. The self is a function of the world as standing reserve, the collection-disposal of accumulated raw material. And the self becomes raw material as well.

It should not be surprising that sustainability and autonomism are two versions of essentially the same mode of "challenging" (in the Heideggerian sense). They are both technological solutions to the dread of human temporality and mortality. Both entail an ideally stable subject that conceives of a natural world as a collection of resources at Man's disposal. The only difference is that the autonomist world is one that emphasizes speed, movement, consumption, and destruction, while the sustainable one stresses consumption, conservation, and recycling. In both cases the standing reserve is there, at the ready; raw materials are there to be used for Man's survival and comfort. Both exist to procure for Man a certain emotional state that is deemed to be morally superior: autonomism supposes a joy in the heedless exercise of individual will ("freedom"), sustainability supposes a dogged contentment through renunciation and the sense of superiority engendered by a virtuous feeling of restraint. In both cases the human self as overweening, protected, permanent jewel is inextricably bound to the destiny of all matter. Bataillean generosity from this perspective is unthinkable. All matter is capable of taking, and holding, beautiful or significant or quantifiable shape; all energy can be refined and concentrated so that it can do "work." The universe wears a frock coat, as Bataille put it in "Formless":

> What [the word "formless"] designates has no rights in any sense and gets itself squashed everywhere, like a spider or an earthworm. In fact,

> for academic men to be happy the universe would have to take shape. All of philosophy has no other goal: it is a matter of giving a frock coat to what is, a mathematical frock coat. (*OC*, 1: 217; *VE*, 31)

I suppose if I were given a choice between the two versions of the world picture, I would pick the sustainable one because it is, well, sort of sustainable—in principle, anyway. In an era of fossil fuel depletion, in any case, we will get sustainability, voluntarily or involuntarily. And certainly planning sustainability in the mode of "powerdown" (Heinberg 2004) is preferable to resource wars and unevenly distributed depletion. Believing in a completely sustainable (unchanging) world is, however, akin to believing in a coherent God. But unless one derives grim satisfaction from renouncing things and contentment with a sense of how much one has had to give up, sustainability as conceived by Newton, for example, is always bound to come out second best. That is why, as long as refined fossil fuels are cheap and no one has to think too much about the future, the suburbs will always win out over, say, sustainable cohousing.[11] An environmentalism that promises *only* a beautiful smallness, or a "prosperous way down," is bound to have little appeal in a culture—and not just the American culture—that values space, movement, and a personal narrative of continuous improvement and freedom (financial, sexual, experiential)—even if those versions of the "tendency to expend" remain in thrall to the self as ultimate signified.

Where does that leave Bataille's future? Recall our analysis of *The Accursed Share* in chapter 2: the Marshall Plan would save the world from nuclear war not because it was the goal of the plan to do so, but because the aftereffect of "spending without return" is the affirmation of a world in which resources can be squandered differently: the alternative is World War III. The world is inadvertently sustained, so to speak, and the glory of spending can go on: this is what constitutes the ethics of "good expenditure." Now of course we can say, from today's perspective, that Bataille was naive, that the "gift-giving" engaged in by the United States under Harry Truman was a cynical attempt to create a bloc favorable to its own economic interests, thereby saving Europe for capitalism and aligning it against the Soviet Union in any future war—and that was probably the case. But Bataille himself was perfectly aware of the really important question: after all, as he himself puts it, "Today Truman would appear to be blindly preparing for the final—and secret—apotheosis" (*OC*, 7: 179; *AS*, 190). *Blindly.* Even if Bataille may have been mistaken about Truman, who after all was giving the gift of oil-powered technological superiority, the larger point he is making remains valid: giving escapes the intentions of its "author." What is important is gift-giving itself, and the good or bad (or selfish) intentions of the giver are virtually irrelevant. What counts, in other words, is *how* one spends, not what one hopes to accomplish by it. Intentionality, with its goals proposed by a limited and biased self, reveals its limits. Derrida noted in his famous controversy with Jean-Luc Marion about gifting that there can never be a real gift because the intentions of the giver can never be completely unselfish.[12] Thus the very idea of the gift is incoherent: a completely unselfish gift could not be given, because it would be entirely without motive. It could not even be designated as a gift. To give is to intentionally hand something over, and as soon as there is intention there is motive. One always hopes to get or accomplish *something*. But, as Marion would counter, there is a gifting that escapes the (inevitable) intentions of the giver and opens another economy and another ethics. This is a gift that, past a certain point, always defies the giver. Of course, one "knows" what one is giving; there are criteria for the evaluation of the gift—but then that knowledge is lost in non-knowledge. The left hand never really knows what the right is doing. Nor does the right necessarily know what it is doing, for that matter.

The ethics of *The Accursed Share:* by giving, instead of spending for war, we inadvertently spare the world and thus make possible ever more giving. Energy is squandered in the production of wealth rather than in nuclear destruction. As we have seen, however, Bataille never adequately distinguishes between modes of spending and modes of energy. Heidegger does: quantified, stockpiled energy has as its corollary a certain objectified subjectivity, a certain model of utility associated both with the object and with the self. Another spending, another "bringing-forth" is that of the ritual object, which (even though Heidegger does not stress it) entails another energy regime: not the hoarding and then the programmed burning-off of quantified energy, but energy release in a ritual that entails the ecstatic and anguished movements of the mortal, material body.

If we read Bataille from a Heideggerian perspective, we can therefore propose another giving, another expenditure. This one too will not, cannot, know what it is doing, but it will be consonant with the post-Sadean conceptions of matter and energy that Bataille develops in his early writings. Bataille's alternative to the standing reserve is virulent, unlike Heidegger's, no doubt because Bataille, following Sade, emphasizes the violence of the energy at play in ritual. Bataille's world is intimate, and through this intimacy it gains a ferocity lacking in Heidegger's cool and calm chalice or windmill (though both represent, in different ways, the lavish expenditure of unproductive energy). Bataille's matter now is certainly not quantified, stockpilable; it is a "circular agitation" that risks, rather than preserves, the self. Through contact with this energy-charged matter, and the nonknowledge inseparable from it, the dominion of the head, of reason, of man's self-certainty, is overthrown: God doubts himself, reveals his truth to be that of atheism; the human opens him- or herself to the other, communicates in eroticism, in the agony of death, of atheistic sacrifice.

Just as in *The Accursed Share,* where the survival of the planet will be the unforeseen, unintended consequence of a gift-giving (energy expenditure) oriented not around a weapons buildup but around a squandering (giveaway) of wealth, so too in the future we can posit sustainability as an unintended aftereffect of a politics of giving. Such a politics would entail not a cult of resource conservation and austere selfhood but, instead, a sacrificial practice of exalted expenditure and irresistible glory. Energy expenditure, fundamental to the human (the human as the greatest burner of energy of all the animals), would be flaunted on the intimate level, that of the body, that of charged filth. The object would not be paraded as something useful, something that fulfills our needs; its virulence would give the lie to all attempts at establishing and guaranteeing the dominion of the imperial self.

One cannot deny the tendency to expend on the part of humans; on the contrary, following Bataille, we can say that this *conscious* tendency to lose is what both ties us to the cataclysmic loss of the universe, of the endless, pointless giving of stars, and at the same time distinguishes us through our awareness, our *savoir,* of what cannot be known (sheer loss). It is vain to try to deny this tendency, to argue that destruction is ultimately somehow useful, that our role here on the planet is necessary, and necessarily stingy. Parsimonious sustainability theory ends only in a cult of the self, jealous in its marshalling of all available resources. We are, on the contrary, gratuitous losers (like any other animal, but more so, and conscious of it), and this is our glory, our pleasure, our death trip, our finitude, our end. If on the other hand we try to substitute a mechanized, quantified, objectified version of expenditure and claim that it addresses all of our needs, our freedom, extravagance will be subordinated to our personal demand, energy will become mere refined power, and we end up running the risk of destroying ourselves on a planet where every atom has been put to work, made to fulfill human goals—and where every usable resource has been pushed to the point of depletion.[13] But most of all, in wasting in this way, engaging in this blind travesty of the tendency to expend, we deny any communication with and through the intimate world, the other torn in erotic ecstasy, the movement of celestial bodies, the agony of God.

For Bataille, in 1949, peace was the unforeseen, unplanned aftereffect of spending without return on a national scale. By expending excess energy through the Marshall Plan, the world was (according to Bataille) spared yet another buildup of weapons. But—and this perhaps was the weakness of Bataille's argument—the Marshall Plan distributed money, the ability to buy manufactured goods, energy stored in products and things. For us today, expenditure entails the eroticized, fragmented object, the monstrous body that moves and contorts and burns off energy in its death-driven dance. Expenditure cannot be mass-produced because in the end it cannot be confused with mechanisms of utility: mass production, mass marketing, mass destruction. All of these involve, are dependent on, and therefore can be identified with a calculation, a planning, a goal orientation that is foreign to expenditure as analyzed by Bataille. At best they afford us a simulacrum of the dangerous pleasure of sacred expenditure (and thus their inevitable triumph over sustainability as austere renunciation). If then we affirm Bataille's expenditure, we affirm an energy regime that burns the body's forces, that contorts, distorts, mutilates the body, and we affirm as well the forces that are undergone rather than controlled and mastered. The energy of these forces spreads by contagion; it cannot be quantified and studied "objectively."[14] Which is not to say that it does not make its effects felt quite literally; the blood-covered voodoo priest in a trance (a photograph reproduced in *Erotism*), L'Abbé C. squirming in agony, and Dirty retching violently (in *Blue of Noon* [1978]) bear witness to this shuddering force.[15]

This energy, however, has little to do with that put to use in a modern industrial economy. This is not to deny that some rational instrumentality is necessary to survival; in order to live, spend, and reproduce, all creatures, and humans above all (because they are conscious of it), marshal their physical forces and spend judiciously. But, as Bataille would remind us, there is always something left over, some excessive disgusting or arousing element, some energy, and it is this that is burned off and that sets us afire.

By separating this loss from industrial postconsumer waste, we inadvertently open the space of a postsustainable world. We no longer associate sustainability with a closed economy of production-consumption; rather, the economy of the world may be rendered sustainable so that the glory of expenditure can be projected into the indefinite future. What is sustained, or hopefully sustained (since absolute sustainability makes no sense), is not a permanent subjectivity that slices and dices and doles out an inert and dangerously depletable (but necessarily static, posthistorical) world; instead, the world is sustained as a fundamentally unplanned aftereffect of the tendency to expend. Unplanned not in the sense that recycling, reuse, and so on, are to be ignored, but in that they are an integral part, inseparable from and a consequence of, a blind spending of the intimate world. The logic of conservation, in other words, is inseparable from expenditure: we conserve in order to spend, gloriously, just as the worker (according to Bataille), unlike the bourgeois, works in order to have money to blow. Thus postsustainability: sustainability not as a definitive knowledge in and as a final, unalterable historical moment, but rather a knowledge as nonknowledge, practice as the end of practice, the affirmation of "nature"—including its fossil fuel energy reserves—that refuses to see it simply as a thing, as a concatenation of energy inputs that need only be managed. Rather, nature is what sustains itself when we sustain ourselves not as conservers but as profligate spenders—not of stockpiled energy, but of the energy of the universe (as Bataille would put it) that courses through our bodies, above us, below us, and hurls us, in anguish, into communication with the violence, the limit, of time. The postsustainable economy is a general economy; beyond the desires and needs of the human "particle," it entails the affirmation of resources conserved and energy spent on a completely different scale. Rejecting mechanized waste, the world offers itself as sacred victim.

The world we face, the world of "Hubbert's peak" (see Deffeyes 2001) and the rapid decline of inert energy resources, is thus, paradoxically, a world full of expendable energy—just as Bataille's austere postwar era was wealthy in a way his contemporaries could not comprehend. The peak of consumption and the revelation of the finitude, the depletion, of the calculable world is the opening of another world of energy expenditure and the opening of a wholly different energy regime. And it is the blowout at the summit of a reason through which society has tried to organize itself. The available energy that allows itself to be "perfected," refined, and that therefore makes possible the performance of the maximum amount of work, in service to the ghostly identity of Man, gives way to another energy, one that cannot simply be retrieved and refined, that defies any EROEI, that does work only by questioning work, that traverses our bodies, transfiguring and "transporting" them. We just need to understand fully what energy expenditure means. Wealth is there to be grasped, recycled, burned, in and on the body, in and through the body's death drive, as a mode of energy inefficiency, in the squandering of time, of effort, of focus.

Postsustainability, Recycling, and the Intimate World

Perhaps these observations can be restated in another mode. A consideration of Agnès Varda's brilliant film, *The Gleaners and I* (2000), will show the ways in which a *future* culture (one that is sustainable only as an after-effect: postsustainable) exuberantly displays its excess, even as rituals of extravagant recycling are reenacted.

Varda's film is an infinitely rich documentary (for want of a better word) that examines all the ways people in the modern world "glean." Gleaning, in both French *(glaner)* and English, is the activity by which people "gather grain left by reapers," and it is also, metaphorically, "to collect (knowledge or information, for example), bit by bit."[16] This metaphorical linkage is important, because Varda starts with the literal senses of gleaning and goes from there to consider all the implications of its metaphorical drift.

We see an example of this near the beginning of the film, where the modern French potato industry is shown. After the mechanized harvesting of a field, literally tons of potatoes are left over; ones that are deformed, too big, too little, too dirty, too mutilated. We see a dump truck dumping them on the side of the field. At this point gleaners arrive and go through the piles, picking out spuds that can still be eaten—either by the gleaners themselves or by guests at the Resto du Cœur (restaurant of the heart), the local charity restaurant where free meals are distributed to those in need. We are invited to consider, by some of the gleaners themselves, the madness of this waste of food when many in the area are malnourished.

But Varda doesn't stop there. Clearly she could write, or film, a dissertation on the inequality of the distribution of food and wealth in France—especially in the contemporary postmodern era, when unemployment and poverty have been unequally distributed, with the young, women, and minorities receiving a disproportionate share.[17] That is implied, of course, but Varda wants to think about more than literal gleaning and its roots in earlier eras, when mass poverty made its practice necessary.[18] Metaphorical gleaning is the taking and reusing of "bits" for purposes for which they may not have been intended. That may include concepts, but also potatoes. Here a potato is not just food, but a heart; a gleaner points out to Varda that many of the rejected spuds are heart shaped, and indeed she picks one up and lovingly films it with her small video camera. The gleaned object suggests by its shape a concept that in turn suggests the restaurant of the heart, but also, of course, love in a more general sense: for the poor who have to glean, for those who take the trouble to find food for the Resto du Cœur, and, in a more general sense, for all others. The recycled object in other words is charged with a powerful but not necessarily useful emotion, communicates that emotion, and triggers in the filmmaker, who is also writing a kind of autobiography through her film, a meditation on the finitude of her own life, her own love, her own aging, her own death.

She brings the potato home; it is given a privileged place in her apartment. Her act of filming the malformed potatoes, which we are shown, is quickly transformed into her filming her own hand: "my project is to film one hand with the other." One hand doesn't necessarily know what the other is doing, but the other films its not-knowing. And what it films is love: the love of Varda's own life, and its passing. Her filmed hand, her left hand, is old, wrinkled; she tells us that through filming she "enters into the horror of it." She gleans her hand, and, like the potato, her hand is charged with metaphor: it represents age; "now the end is near," she tells us. No doubt her hand also represents love, which, as we know, is a form of gleaning; the gleaners use their hands to glean; Varda uses her camera, holds it in her hand, points it at her hand, to glean images. She gleans the image of her loving hand, the gleaning hand, dying, indicating the limit.

The gleaned object dies; it is death. She keeps the heart-shaped potato, and later we see it, wrinkled, decaying, disgusting, but still alive, sending out grotesque "eyes" (buds). It is still heart shaped.

There is something of the surrealist technique in all this, the metaphorical transfer of the image of one object into another (Salvador Dalí was a master of this) conveying a world charged with emotion, with horror. The gleaned object is not just matter to be stored, sliced and diced, used up, wasted, thrown out; it is, as Bataille would say, an intimate world, one that cannot be mastered, rendered neutral and useful in some task. It is not a means to an end, but an end; the end is its force, its power of suggestion, its erotic, totemic sway, its social (de)centrality. It is "formless": it cannot, will not, contribute to elevated concepts and definitive understanding. But we must not forget that it is, precisely, recycled: it is recuperated, reused, in its very disposal. Recycling and trashing are inseparable, and trashing is an affair of the heart. The object's charge is thus inseparable from the (metaphoric) transfer that is its dumping, its expenditure: its charge is not only its end, but its linkage of the eroticized death drive (recall Varda's hands—the gift of death), charitable communication with another (the giving of food through the heart), and ludic, aesthetic practice (the gift of Varda's film itself). The charge of the object is this instantaneous movement, the scar of its past life, the impossibility of its appropriation in the elaboration of a future end. Through recycling, what may have been *wasted* is *expended.*

The recycled object is also violently riven by the trace, the scar of its former appropriation, the mark of its imminent and immanent death. The object is autobiographical; it tells its own story in the marks of its past life. There are a number of other artists in the film besides Varda, and they speak of the salvaged objects they use in their work. One of them, VR2000, maintains that discarded objects are "still living," that they "had a life," and that they "deserve another." He cruises the neighborhood at night on his bike, carefully scanning piles of trash. In VR2000's works, assemblages, and paintings, the objects bear witness to their own status as refuse, as rejected matter; they tell their own story, that is their second life, and in that way they confront the viewer, demanding her attention, demanding perhaps to know why they have been so easily trashed. An old Russian mason, Bodan Litnanski, creates towers using discarded objects as his bricks: grotesquely damaged dolls bear witness to lost love, sick fetishism, the aging of their former owners, the ravages of weather and time. Litnanski no doubt would have us love his art the way a fetishist loves a shoe.

Finally, Varda herself comes up with perhaps the ultimate gleaned object: a clock face set in a transparent base, without hands, a hole in the middle. Varda loves this fragment, sets it up in her apartment, passes behind it, looking at us. Here we have the recycled object in all its glory: useless, perhaps, in its conventional destiny, which is to quantify time, allowing us to "budget" it. It now indicates its distance from the smart use of time; it shows itself as end, without higher end; it displays this moment through which Varda moves. It is charged with the glory of lost time, of the loss of all desire to maintain time, of death as the most fundamental loss of time. It is the eroticized moment, the "now" divorced from all movement. Its mechanism is missing: we supply the movement of time through our own bodily power by looking at it, following its empty dial with our eyes. Varda looks through it and sees her own reflection, her own autobiography of the instant (the now, love, death) in the lens of the camera, in our gaze. Its charge is that of utopian revolt, of the workers celebrated by Bataille who find their meaning in throwing away their wealth, of the young people in Varda's film who defiantly rummage through trash bins for food and who openly revolt against the law when a supermarket manager pours bleach on the food to render it inedible. It is found art, like the composite painting—gleaned from other paintings of gleaners—that Varda gleans from a secondhand art and furniture emporium ("Discoveries *[trouvailles],*" the sign proclaims).

If there is to be a postsustainable world, it will open itself as the aftereffect of gleaning: of the charged object, the charged body, the collision of past and future in and through death. It is a realm where there is a convergence of responsible recycling, defiant ritual, the sacrificial destruction of use and meaning, and social commitment beyond the narrow desires of the self. Community is an aftereffect of such postsustainable generosity. Recycling, as in Varda's film, will entail not just a practical reuse of a salvaged standing reserve (reused material: plastic, steel, paper, etc.) but, more profoundly, a kind of erotic reinvestment and disinvestment, in which the object takes on a meaning that defeats our demand that it be a simple tool, a simple means to the end of status, individuality, comfort. Food itself (what today we call garbage) will be a sacrament of the death of God, barely edible trash[19] reborn as a symbolic crux of life and death, meaning and non-knowledge. This will be a moment in which the aesthetic—many of Varda's gleaners are artists—intersects with the religious (the orgiastic, the sacred) and the practical (the only "stuff" available is what can be recycled). In a "depleted" (postcarbon) world where, eventually, everything will be junk, and in which there will be little official energy to create new junk, orgiastic recycling will tear us from our projects and project us into communication with others, with the void. The tossed object will be consumed in an intimate world. "Bits" will be everywhere, the traces of the lost world of seemingly infinite standing reserves; they will be reused and they will command their reuse, talking to us, signifying our own loss and our own glory in reusing them to live, to play, to do nothing. "Bits," like walking or cycling, will both enable us to go on living and, as modes of squandering (burn-off of time, of effort, of sanity), constellate an intimate world.

To paraphrase Bataille, only a madman would see such possibilities in the recycling of a few bits of refuse. "I am that madman" (*OC,* 7: 179; *AS,* 197, note 22).

6 The Atheological Text

Ecology, Law, and the Collapse of Literalism

In this chapter I continue with the set of problems I examined in chapter 3: Bataille's "religion" of the death of Man, and God, in and through his counter-Book, the *Summa Atheologica.* Any thinking of sacrifice in the era of depletion must pass through a theory, and hence a "religion," of ecology. Ecology alone allows us to think the future as a play of forces and resources, modes of conservation and expenditure. Ecology inevitably entails thinking about sustainability or, in the case of Bataille, postsustainability. A number of writers in recent years have attempted to pose the problems of ecology and sustainability in a religious context. Predictably, those theories have entailed a cult based either on Man or on God; two options expressly critiqued by Bataille. It is up to us, his readers, to think about the passage from ecoreligions based on these unavoidable but ultimately hollow figures to one based on a Bataillean model of an expenditure of those terms, put forward, and lost, in a counter-Book.

Indeed, as resources grow more scarce, a consoling and judging God looms ever larger: if indeed a solar economy implies a feudal economy, oppressive and literal-minded religiosity may very well lurk on the horizon.[1] It is all the more imperative, then, to anticipate another, solar religion, not of hellfire and obedience but of the fall, the irreparable loss, of God.

Ecoreligion, the Book, and Sade

Perhaps no article has haunted contemporary ecological thought as thoroughly as has Lynn White Jr.'s essay "The Historical Roots of Our Ecologic Crisis," which first appeared in *Science* in 1967. White's essay, though seemingly modest in scope and ambition, has nevertheless had a profound impact on all debates concerning ecology and the exploitation of resources.

Any thinking of ecology, religion, and law in the twenty-first century must inevitably confront White's straightforward thesis. It is much easier to ignore the questions he poses or to willfully misconstrue them than it is to refute them effectively.

White posits the need for an ecological history, that is, the history of ecological change, which is, he argues, "still so rudimentary that we know little about what really happened, or what the results were" (1968, 76) in the interaction between human activity and ecological degradation. Human environmental impact on species populations, the subsequent changes in the quality of life and the course of history—these problems, according to White, "have [never] been asked, let alone answered."

This simple paragraph has, of course, spawned a whole school of historical analysis, epitomized by William Cronon's great study of the rise of Chicago and the resulting transformation of the social and natural ecology of the Midwest, *Nature's Metropolis* (1991). But no doubt the greatest influence of White's little essay has been in the area of ecology studies. White's thesis, in a nutshell, is that the current "ecologic crisis" is the result of cultural attitudes inculcated through the teaching of Judeo-Christian religion. Nature had always been represented by religions, but up until Judaism it was seen as somehow independent, animated by autonomous spirits:

> In Antiquity every tree, every spring, every stream, every hill had its own *genius loci,* its guardian spirit.... Before one cut a tree, mined a mountain, or dammed a brook, it was important to placate the spirit in charge of that particular situation, and to keep it placated. (White 1968, 86)

Judaism, and after it Christianity, reversed this relation so that now nature was no more than the material that constituted it. Not only did this religious tradition insist that it was "God's will that man exploit nature for his proper ends"; that exploitation was made possible by the emptying out of all particularity, all life, from the natural object: "Christianity made it possible to exploit nature in a mood of indifference to the feelings of natural objects" (White 1968, 86).

Now one could certainly argue that White's thesis bears some similarity to Heidegger's in "The Question concerning Technology," while no doubt lacking the master's subtlety. While White is not concerned with *physis* and the various modes of "revealing," his critique nevertheless has a real power since it points to a single text, and a single section within a text, as the cause of the problem. For White, it is Genesis that gives us to understand that God has created nature *for us.* Adam names the animals, gives them their particularity, gives them, as Hegel and Blanchot argue, their mortality, since the name is the possibility of abstraction, the possibility of constituting something by separating it from the necessity of its undifferentiated existence.[2] The natural world as configured through language by humans is already dead, already merely a resource at our disposal, and it has been created precisely to fulfill that role: God would have it no other way. Outside of Man, his desires and needs, his objectifying speech act, nature would have no existence. God created nature so that it could be delineated through naming and put to use.

White therefore argues that there is a simple source for misunderstanding man's role in nature and the very being of nature itself: a certain religion, one that links monotheism, the Book, and the subordination of nature to the human. For the Book is implied as well and is central to the drama White outlines: without the Book, without the single, unambiguous story told in Genesis, there might be room for "guardian spirits" to reassert themselves. But the written text warns us, attempting to prevent this: there is only God, his Word, and creation, and creation is for Man. By presenting the problem so unambiguously, White invites us to undertake a straightforward project: reconceiving a single religious tradition, rejecting a large portion of it, and embracing, as we learn at the end of White's essay, one figure who runs against it and who still can offer us some hope: St. Francis of Assisi.

This last suggestion has been largely ignored by virtually all writers who have come after White: they tend to see only a blanket condemnation. Perhaps St. Francis's offer of a viable alternative to the Christian tradition is ignored because St. Francis himself, as White admits, was almost immediately reappropriated by orthodoxy. Not a very auspicious beginning for an attempt at wresting Christianity away from its tradition of instrumentalizing and objectifying nature.

Yet ecological theorists just can't seem to leave religion alone. It offers too much; we have already seen, in the preceding chapter, how a morality based on Christian "simplicity" can be allied with an austere ecological morality (sustainability). Beyond this, critics would like to appropriate a sense of wonder or the sacred for ecological awareness.

Gary Gardner, for example, in his book *Invoking the Spirit: Religion and Spirituality in the Quest for a Sustainable World* (2002), points out the many ways that religions and religious organizations can be harnessed in the effort to further the transition to sustainability. He has, for obvious reasons, little to say about Lynn White's thesis; rather, he notes a number of aspects of all religions that would make them obvious vehicles for organizing people to struggle for an ecologically sensible life practice. He notes that religions typically "provide meaning" in ways that growth societies cannot; further, this "capacity to provide meaning" is

> rooted deep in the human psyche. . . . This capacity is often expressed through symbols, rituals, myths and other practices that work at the level of affect. These speak to us from a primal place, a place where we "know" in a subconscious way. Ritual, for example . . . is a deep form of communication that is tapped by both religious and secular leaders. (13)

Gardner goes on to note that rituals that tie together populations are a feature of traditional societies, and that such rituals, with their worldview-altering capacity, could easily be implemented in activities that foster a "green" attitude on the part of the population. Clearly, such rituals are tied to affect, and our sense of wonder and rapture in nature is inseparable from a religious orientation.

Religion has also, according to Gardner, traditionally exhorted believers to "avoid preoccupation with wealth and materialism" (47–48); religions, which have "centuries of experience reading their central tenets in the light of contemporary realities" (49), could therefore be expected to incorporate sustainability as a tenet of belief. Gardner has a number of suggestions (41–43) for congregations that would like to incorporate green activities in their ministries, such as the Episcopal Church's Episcopal Power and Light (EP&L), a "ministry that promotes green energy and energy efficiency" (42). Indeed, one can easily imagine an energized congregation working, as part of its "ritual," to implement, on a community-wide scale, sustainable energy practices, organic farming activities, and so on.

That, then, is one way of refuting Lynn White—or perhaps of continuing St. Francis's ministry, of which White would no doubt approve. The larger problem of religion as vehicle for the vision of creation as anthropomorphic domain—a garden created by God for Man, with Man in the image of God, and all other beings there merely for Man's use—is conveniently sidestepped. From garden as Man's domain we pass quickly to religion as antimaterialism, which presumably entails a light treading on the earth.

But this is still an overwhelmingly anthropocentric vision. It is a sympathetic one, to be sure, and one can think of worse things to do with one's time than engaging in the ecoreligious ministry advocated by Gardner. But one has the sense that there is a certain forcing of the religious tradition going on in most ecoreligious writings, along with an unconscious but stubborn clinging to the vision of Man at the center that White would see as the root of the problem in the first place. This combo of consciously rewriting a (or the) religious tradition (or selectively rereading it), along with a continued insistence on Man's central role, has some unhappy consequences.

Mary Evelyn Tucker, a leading authority on ecoreligion, writes, in her book *Worldly Wonder: Religions Enter Their Ecological Phase* (2003), of the need for religions to get with it, so to speak, and do the heavy lifting necessary to change the consciousness of the citizens of the planet:

> Although our deleterious role as humans is becoming clearer, so, too, are various efforts emerging to mitigate the loss of species, restore ecosystems, . . . and preserve natural resources for future generations. The question for religious traditions, then, is how can they assist these processes and encourage humans to become a healing presence on the planet. (9)

Tucker's tone is more pragmatic than Gardner's. She wants religion to get to work to make the new ecoconsciousness a reality. Religion is to be put to the test, given a task; its role is ultimately a practical one. Moreover, religion's job in this view is to resituate Man on earth: "What is humankind in relation to thirteen billion years of Earth history? How can we foster the stability and integrity of life processes?" (8). This line of questioning in fact answers itself: humankind is central to "Earth history," else there would be little need for us to "foster the stability and integrity of life processes" (8). (One cannot avoid thinking: those processes hardly need us to foster and stabilize them; they could do it quite well, thank you, without us.) We have a "niche," religion will help us find it (7), and, it is assumed, we have to be here to "[create] viable modes of religious life beneficial not simply for humans but for the whole Earth community" (10). To be sure, Tucker stresses "human-earth" relations, rather than "God-human relations and human-human relations" (9), and in that way she can be said to be taking into account White's critique; it is not just us and God anymore, with Earth as resource pile. We are "universe beings" (11). Tucker nevertheless stresses, implicitly, that Man not only has a niche but is in some way necessary to nature, else we would hardly have to worry about *doing something* to "foster" sustainability. Earth is, in other words, a project, and only Man can carry out a project. Without Man Earth will somehow be lost—and we are back to a version of Genesis, the creation of the world for Man, and (another version of that doctrine) Man as necessary steward of the earth.

Along with this mission for Man comes the separation between matter and spirit. Indeed, Tucker writes that the religions of the world "are awakening to a renewed appreciation of matter as a vessel for the sacred" (9). Matter here is the very stuff of which natural beings and processes are constituted; but matter above all contains the spirit to which religions have access. Religions, in turn, are strictly a human affair; animals and plants do not worship. By stressing this opposition, Tucker reaffirms the superiority of humans who have the responsibility not only of saving nature, but of appreciating (experiencing? venerating?) spirit through matter. In Tucker, unlike Gardner, spirituality is not only antimaterialism (rejecting the gaudy fossil fuel lifestyle) but the sanction of the necessity of the human in the cultivation of an elevated, spiritualized matter.

The irony, of course, is that by attempting to answer Lynn White's critique—religious anthropocentrism breeds ecocide—Tucker only reaffirms anthropocentrism, along with its twin, the matter/spirit dyad. Nature once again is valorized primarily as a vehicle, an environment, for the human. Man is needed, as guardian, as steward, as subject of spirit. Nature without him, and perforce without God, is not only mere matter, but is somehow unsaved—else why would Man be needed to foster and sustain? Tucker, in effect, overturns White's model: now it is not ecocide generated by the subordination of nature to Man; it is nature that is, or at least that can be, saved by that subordination.

The larger problem is maintaining the pretense that religious experience is something more than an expedient to the preservation of nature. If we reconfigure religion only to "foster and sustain," then to what extent can we continue even to believe in the independent power and validity of religion? Doesn't it simply become one more tool, suited to the accomplishment of a task? Tucker holds up one element of belief that transcends the various doctrines: spirit. But if spirit is little more than a code word for

Man (it is what differentiates Man), can we see ecoreligion as anything more than a humanism, a tool that Man would use to survive, to further his own ends? If this is the case, haven't we simply inverted the model analyzed by White? Instead of nature as a commodity given to Man by God, for Man's use, God is now a commodity given by Man to Himself in order to better use (by justifying the protection of) nature.

If this is the case, the faithful may very well have a harder time believing in God, and the very notion of an ecoreligion will succumb to bad faith. We see this, I think, quite clearly in Bill McKibben's concluding remarks in an issue of the journal *Daedalus,* edited by Mary Evelyn Tucker and John A. Grim, titled *Religion and Ecology: Can the Climate Change?* This issue contains, among other things, an overview of various world religions (Islam, Buddhism, Hinduism, Confucianism, Christianity, among others), with considerations of how their traditions in fact can be read to foster ecological awareness rather than to discourage it. McKibben, the author of the influential book *The End of Nature,* contributes a short essay, "Where Do We Go from Here?" (2001) which draws conclusions based on a reading of the volume's essays.[3]

McKibben confronts an obvious problem: how to energize resistance to ecological damage and decline. Like Tucker, he sees great hope in the marshalling of religious faith and institutions in this struggle. And again, he recognizes, if only implicitly, that what had passed as the great Christian monotheistic tradition is not entirely appropriate for ecostruggle. Yet the tradition has something to offer. It is only a question of rereading it. McKibben writes:

> For many Christians, a profound understanding of the Jewish story of Exodus as an allegory of liberation followed, not preceded, Rosa Parks's decision to stay in the front seat of a Montgomery bus. She sat there out of some intuitive sense of right and wrong, of frustration and hope. But as the churches took up the cause, they searched more deeply through their traditions, and certain verses came to new and real life; certain themes emerged. (2001, 302)

In other words, reading *follows* action; we do what we believe right and just, and return to read religious texts, the Bible, in such a way that our "intuitive" actions take on a larger, religious sense. This, presumably, serves to justify future actions and gives us a larger, more coherent, common purpose (we act in and through religious belief, religious texts).

McKibben describes here not so much a mode of religious reading as a secular one. He may very well be right—I am not challenging the legitimacy of his argument—but the process of reading through consensus characterizes not only the reading of religious texts but *all* texts, over time. Reader-response critics, such as Stanley Fish, note that the meaning of literary texts changes over time, and does so through the changing horizons of expectation and meaning determined by interpretive communities. The text might not change, but what we ask of it, what we find in it, what it gives us, is indeed transformed over time. Different communities bring different demands to texts and take away different meanings; no meaning, so long as it is based on textual evidence, is wrong; there is no one absolute meaning, since there is no reading possible outside of the horizons of a community.[4]

This is not relativism but rather a way of conceiving the grounding of meaning in and through socially conditioned activity. It is markedly at variance with a literalist mode of reading, which assumes, on some level, that God has determined an absolute (eternal) message for the salvation of the faithful and written a text susceptible to a single interpretation (be that one of literal reading, allegorical reading, or theological codification). The message McKibben derives from a reading of the Bible would no doubt be markedly at variance with the common interpretation of the Genesis story (critiqued by White) that the creation of the earth, and the creation of all living things, was exclusively for the use of Man (in his worship of God).

> Only our religious institutions, among the mainstream organizations of Western, Asian, and indigenous societies, can say with real conviction, and with any chance of an audience, that there is some point to life beyond accumulation. In the past, that vision was expressed purely in spiritual and esthetic terms; now it has also acquired a deeply *practical* urgency. Those in monks' habits are joined by scientists in white coats, and they're saying the same few things: Simplicity. Community. Restraint. (McKibben 2001, 304; italics added)

Note that it is only now that the spiritual message has acquired urgency (once again, the message is "simplicity"): now the religious message has "practical urgency" because, presumably, we are destroying ourselves along with our planet. Religion is profoundly practical, and its values will help us turn away from the death culture of "accumulation." There's a use value of religion, then, and it is survival. That is why we turn to it, and in the end our doing so enables ecological concern to save religion, just as religion will have saved the ecology movement. "Ecology may rescue religion at least as much as the other way around. By offering a persuasive and *practical* reason to resist the endless obliterating spread of consumerism, it makes of Creation a flag around which to rally" (305; italics added).

Religion is a flag, and we know what purpose flags serve: in battle, troops rally around them and fight to the death, even if the flag itself is meaningless, a mere piece of cloth. The flag is a totem that serves to unify a community, a kind of speech act through which the community calls itself into being and acts.[5]

However, again we have the problem of bad faith. Religion owes its very power to its absoluteness: God is not made by us; we are made by God. The strictures of religion are meaningful and valid because we believe they come from a higher authority. Of course one can argue that secular documents are objects of belief and reverence (the U.S. Constitution, for example), and that they are in a sense "obeyed." But McKibben wants more than the secular authority of the Constitution: he wants the absolute power, the limitless authority, of God. That authority, the authority of the sacred, not the reasoned laws of man, will tear us away from inherently immoral activities (such as accumulation), ones from which, apparently, we cannot be weaned by reasoned arguments alone. But the only way McKibben can get the divine authority he needs for ecoreligion is by consciously tailoring the religious tradition. What authority will such religion have after it results from the obviously pragmatic (hence the emphasis on "practical") strategies of ecoauthorities such as McKibben himself? "There's a real opportunity here, one not yet fully tried," McKibben tells us (304), and we can't help but be swept up in his enthusiasm, at least until we stop short, feeling a bit guilty, before this glow of sheer boosterism.

Perhaps this is why religion and science can lie down together so easily: the message of religion proposed by McKibben has already been confected with *that* practical goal in mind. Now all we have to do is forget about all that and go to church and worship.

We might consider for a moment the role of the Book (the Bible) that McKibben proposes. The Book must be absolute, have the value of overarching authority, but at the same time it is wholly contingent upon readings carried out at specific historical conjunctures by the community of faith (of readers). (Indeed, the process of reading that McKibben affirms may very well have been used all along—but McKibben is cynical—or practical—enough to affirm it openly.) It is those acts of reading, based on recent events, on the needs of the moment, that are exercises of the will—they determine, in short, how the Book will be read and what the larger values of the religion are. Don't like a God who creates the plant and animal world as resources for man? No problem—here's another God, in another passage, one who has given us, and the animals, a quite different position in creation.

In other words, through our own reading, we have written the Book. And yet its *practical* value is that it is absolute: it is the word of God. The Book is not just another human document. We accept its power, its status, but at the same time, as practical beings, we cannot ignore that we have (re)written the Book with a purpose in mind: saving the world for Man. We are happy to have it both ways: we write the Book we want, then accept its Word as usefully definitive. And it can never be a question just of saving the world, because we could literally do that by doing nothing or, more accurately, by dying off. "Saving" the world as a human activity necessarily still places Man at the center: we are saving it for ourselves, so that we can flourish in an ecology that is (relatively) healthy. We are saving it after we have nearly destroyed it.

The Book, in other words, situates our subjectivity as liars; we ourselves have authored a Book that we then take to be absolute, for our own good. The Book is supremely powerful only at the cost of our recognition that we ourselves are weak, hypocritical: why indeed do we need a sacred text to do for us what we cannot do on our own? The Book's coldness, its power, derives first from what it would do for us, but ultimately from its merciless revelation of our own duplicity. It is as strong as we are weak, or, like God as analyzed by Bataille, its very strength, its authority, is our weakness, our need for some transcendent guarantee.

We are not that far, in fact, from Sade's Book in which the absolute, cold Book was both an obvious concoction resulting from human desires (a desire for an economy based on, as we have seen, a rigorous, orgiastic recycling) and at the same time an absolute of power before which we, as readers, are revealed as weak fantasists with quibbling, moralistic doubts: in other words, as victims, sharing the fictionality of the ultimate weakling and victim, God himself.

From anti-Bible (Sade) to postmodern Bible (McKibben), from cold and cruel black star of immorality to the pick and choose gospel of current Christianity—we have not gone that far. Both texts situate the author/

reader as powerful: Sade's reader imagines himself or herself a great "libertine," McKibben's Christian sees himself or herself as the Human that saves the world, the subjectivity that *is* the world. But in both cases, as the Book becomes more powerful, the reader is resituated: now she or he is a creature of bad faith, wracked by doubt, a victim of the monster he or she has created through reading. The Book stares back, absolute, but now it is the human that is contingent, false, precisely because the Book itself is the very pinnacle of falseness, of fiction. The Book is now invincible in its silent accusation of Man. Human "libertine" desire, humanized ecological harmony, both can be purchased only at the cost of creating something dependent on evident fictions, needing the fabulations of a subjectivity that situates itself as omnipotent, but cannot help but know itself in the falsehood of that position. In the knowledge of one's own fictionality, it is now the arbitrarily created Book that takes the upper hand, reveals itself and revels in its final power. It is the Book that dispenses that knowledge, the knowledge of Man's own good and evil, the inescapability of one's lies, the mendacity of one's "virtue." This is the final and ultimate truth of the Book.

The Literalist Text and Its Absence

If humanist religion—epitomized by ecoreligion—is a testament to Man's bad faith, the solution must lie in another faith and another Book. If we dethrone Man by arguing for the supremacy of God, then the text must be, is, authored by God and not Man. Literalism is the simple realization that a humanist Christianity (for example) is an oxymoron. Man is created in God's image, not vice versa. And the truth of that image is given to us, without any possible ambiguity, in the Book. Human life is sacred not in and of itself, not because it is human, but because it is in God's image. The sacred is divine, not human. And so on.

This realization of the antianthropocentric nature of revealed religion will necessarily determine how one sees, and perhaps reads, the Book. In recent years Roman Catholics have been struggling with this problem: what is human, what is divine, in the holy scriptures? The official (Vatican) doctrine, trying rather ineffectually to steer between the extremes of relativism and literalism, proposes a two-track reading of scripture. In the words of Peter Williamson, the explicator of the Vatican's recent statement, *The Interpretation of the Bible in the Church* (2001):

> By distinguishing the divine message in Scripture from its historical conditioning, exegetes help systematic theologians avoid the extremes of fundamentalism, which confuses the human and divine elements, and dualism, which completely separates a doctrinal truth from its linguistic expression. (281)

In other words, the Roman Catholic exegetical ideal is the melding of critical reading, which carries out historical and linguistic analysis of Scripture, with theology, which recognizes Scripture as God's Word. The Bible in this view can be analyzed as a text written at a specific time, by humans, that nevertheless is doctrinally true. Catholic exegesis never assumes that every single word in the Bible is simply true, authored directly by God, and therefore not in need of analysis; nor does it assume that the Bible is simply literature, one work of human writing among others. Clearly, the ecoreligion position, holding that we can derive meanings for our age from a rereading of the Bible, errs in this latter direction; if the Bible can always cough up new meanings, it risks being only provisionally true, its meaning of value only relative to a given historical or political conjunction and determined by whoever is reading it—a postmodern Bible, in effect.

The other extreme, literalism, is just as tempting. One can argue that the Catholic Church can afford to make this distinction because it itself is (or at least sees itself as) already universal, absolutely true and valid. The problem of reading is resolved on a higher level: we can say with certainty that there is a median between the two extremes of reading because the Church itself, given its privileged status (one Holy, Catholic, Apostolic Church) incarnates Truth. No need to argue how the mediation can take place when the Pope is the direct inheritor of the authority of St. Peter, the stand-in for Jesus, who himself miraculously incarnated the conjunction of human and divine, contingent and eternal.

Without the infallibility of the pope and the theology of the Church, one is therefore thrown back on one of the two options Williamson mentions. For a literalist, the absence of a supreme temporal authority (a Church or Caliphate) means that the Book itself is supreme, literally the Word of God. It is not only metaphorically and doctrinally true, it is literally true. The Bible and the Koran therefore come to take the place of God on Earth, and they take the place of authoritative commentators and/or a universal Church as well.

What then of reading? David Zeidan, in his book on Christian and Muslim fundamentalism, mentions Bob Jones II's position on biblical analysis and commentary:

> Chancellor of the separatist Baptist Bob Jones University, [Jones] advocates the method of the Bible interpreting itself. He claims that "the Bible explains itself and you can only interpret an obscure passage honestly when you do so in the light of clear passages." (2003, 149)

Zeidan goes on to cite Jones, who criticizes overreliance on commentaries:

> The important thing [writes Jones] is to read the Book, not books about the Book. Preaching should be soundly Biblical, not coldly and technically theological. Men need the clear proclamation of the great Bible fundamentals—about which there can be no differences of interpretation, for these are positively and explicitly stated. (Bob Jones II, cited in Zeidan 2003, 149)

With the Word of God as truth, free of commentary, fundamentalism necessarily entails "the Bible explaining itself." Man once again is displaced by the Book; reading becomes a movement through which Scriptures can be read without Man. Preaching, one can infer from what Jones says, is merely the act of relaying the reading the Bible does of itself to others. Since the Book is definitive Truth, and reading is not a contingent act but a sheer repetition carried out without, or beyond, human intervention, we are back in the domain charted by Kojève: reading as sheer repetition of an absolute and unchanging Book.

Integral to the fundamentalist conception of the Book is the idea that it is more than mere inspiration for believers or toolbox for the bricolage of a humanly satisfying ethics. To the contrary: it is up to humans to live up to the standards put forward by the Book, the Law of the Book. With the disappearance of Man as the author of the Book and its orientation as the literally true Word of God necessarily comes the position that the Book is Law—since the Book is now the source of motivation that previously had belonged to Man. An authorless, readerless Book, Scripture is the triumph of the Law. One thinks of Blanchot's remark on the Law, where, paraphrasing Sade, he states, "Against the Law, which everywhere constrains me, there is no recourse: the Law wants me always to be deprived of myself, always without passion, that is to say mediocre, and before long stupid" (1992, 225).

Of course Biblical "authorities" like Jones would reject any suggestion that the absolute Law renders one mediocre, but it is true that the Law, imposed from without, substitutes for any autonomous judgment on the part of Man. Thus it is not surprising that the fundamentalist Bible explains itself, since in a world in which the Book is supreme, beyond the human, any act of human reading or human interpretation is pointless. The meaning of Scripture, the meaning of an act, all is given once and for all, beyond debate or cognition, as Law. Morality is therefore nothing more than rote repetition and application.[6]

Or is it even a question of application? Is there a need to use the text as a Law, to restrain or somehow configure human behavior? If the text reads itself, is there necessarily any contact, any traction, between the text and, say, human justice on earth? Does the Law of the text have any say in human activity, human justice, beyond the injunction to respect or revere the text, the Scripture of God itself?

This is certainly an objection that Muslim literalists bring to the Christian tradition. In their eyes, there has been a radical disconnect from the very first between the world and the word of (the Judeo-Christian) God. Sayyid Qutb, a leading modern exponent of a return to a strict conception of the Shari'ah (fundamental Muslim law) criticizes the Christian tradition:

> At some periods Christianity reached a high level of spiritual purification, material renunciation and emotional forbearance. It fulfilled its duty to this side of human spiritual life insofar as it is possible for spiritual teachings lacking a Shari'ah to raise the spirit and elevate the emotions, purify the heart and soul, curb the instincts and dominate their demands, while aiming with sacred longing at the world of ideals and imagination, leaving society to be ordered by the earthly laws of the state in the practical everyday world. (1978, 3)

In other words, the Christian tradition neglected the law of the everyday world, leaving it to corruptible secular authorities. There may be more, however, to this split between divine and secular law. Divine law, with the divine text, would seem to be absolute, but even Qutb recognizes freedom of conscience in the social makeup of the Ummah, the worldwide and eventually posthistorical Islamic community he anticipates (139).[7] Freedom would imply disbelief, pockets of nonapplication or purely empty obedience to the universal Law—in other words, a refusal to read the divine text and believe. Without free will, without the freedom to refuse, the divine law, the Shari'ah, is no longer a law at all but a purely mechanical set of functions, inseparable from natural laws. If that indeed were the case, there would be no point in reading (or repeating) the law, the text as a God-given part of a system of religious belief. Thus a paradox: the divine law, applicable to all aspects of life, and not only to the spiritual, inward life, itself generates a pocket, a space of resistance, necessary to its function as Law.

Nevertheless, Qutb's point concerning Christianity is well taken even if the "pocket" of secular law may not be that different, in the end, from the "pocket" implied in (Qutb's version of) Islamic law with its (at least theoretical) insistence on free will. The divine Law in Christianity, in its very divinity, carves out a space where it, as Law, has no business: what results is a split between the spiritual and secular worlds and between spiritual and secular law.

No doubt one of the fields where this connection, and split, between divine and secular law is felt most acutely in literalist Christianity is the area of ecological responsibility. Evangelical ecological critics respond to Lynn White Jr.'s critique by citing biblical verse that man is the steward of God's creation and thus responsible for its health. There is more to the Genesis story than God creating the earth as nothing more than an easily exploited resource base for humans. Christianity in the Evangelical view is the ultimate ecological religion because God has ordered Man to take care of His creation. In that sense they are perhaps not so far from McKibben and others who, as we have seen, argue that the Bible should be reinterpreted from an ecological standpoint. Other Evangelical critics, however, cite Biblical verse to refute this effort, in a very interesting way. E. Calvin Beisner, for example, in his cogent book, *Where Garden Meets Wilderness,* notes that Jeremiah 2:7–8 is often cited by Evangelicals as proof that God punishes man for the defilement of the land (1997, 46). Beisner's point, however, is that the biblical passage in question does not invoke the punishment of man by God for man's environmental sins; on the contrary, God *caused* the land to become sterile because men were not faithful to God. Beisner comments: "What sins 'defiled the land'? Not poor environmental practices, but idolatry and infidelity to the covenant" (47).

Beisner, implicitly at least, makes the argument that the most ecofriendly activity one can engage in is worshipping God sincerely and, no doubt, reading the Bible in its literal sense. Only in this way will the land be blessed and fertile.

> Notice that when Israel had been faithful, God had blessed it by bringing it out of foreign oppression in Egypt, leading it safely through a barren wilderness, and settling it in "a fertile land to eat its fruit and rich produce" (2:6–7). But now that Israel has turned in betrayal from the covenant, God is judging it by putting it under foreign oppression again and making its land infertile. (Beisner 1997, 47)

There is no reason to believe (nor does Beisner suggest) that today is any different from biblical times. Ecological devastation is a direct result of disbelief and idolatry.

In this view, the biblical text as Law reigns supreme, but it has little to do with the actual management of the earth—in other words, it is neutral as to human affairs except when those affairs involve a lack of recognition of (respect for) God. The text as Law means that the worship of God is central: man and his acts, as they pertain to his mere survival, are not. While relativism implied a human-centered religion and a text inseparable from its contingent (situational) readings, fundamentalism entails an absolute text that imposes itself as Law, with man subordinate to and dependent on God. Evangelical literalism is not humanism, and, we can conclude from Beisner, it does not easily lend itself to revision as ecoreligion. Concerning Lynn White Jr.'s argument, Beisner remarks:

> Truly Biblical Christianity denies precisely what White attributed to Christianity: the view that "God planned all of [creation] explicitly for man's benefit and rule: no item in the physical creation had any purpose save to serve man's purposes." Indeed, rather than being "the most anthropocentric religion the world has seen," as White charged, Biblical Christianity is God-centered. For Biblical Christianity tells us that God created all things for His pleasure (Rev. 4:11) and that the heavens declare not the glory of man but the glory of God (Ps. 19:1). (132)

Beisner stresses that God gave man "authority over everything on earth (Gen. 1:26–28)" and in that sense man "also exercises some authority over whatever it is that he stewards" (133). The result, perhaps as Qutb might have foreseen, is a split between the results of that authority and the divine strictures in the text. Since the text is God-centered, not man-centered, it has little concern with *how* man's authority is exercised; it is concerned with how God is worshipped. One can see quite clearly the modern implications of this version of textual authority. Beisner argues, in a chapter titled "The Problem of Environmental Misinformation" (1997, 59–79) that oil depletion, resource degradation, ozone layer disappearance, acid rain, increased carbon dioxide in the atmosphere, and so on are nonissues. Oil depletion, for example, has been predicted since 1866 and has always proved to be a false alarm. Thus, by implication, it always will be a false alarm (64).[8] The impression one carries away from this pseudo-scientific exposition, cheek by jowl with biblical exegesis, is that we do not need to worry about what is happening to the environment because it will take care of itself *if we are good Christians.* Ecosystem collapse is nothing more than a symptom of disbelief, of failure to read the Bible faithfully. God put the earth there for us to use and to use in his praise, not for our own purposes. Therefore if we use it while remaining believers, we have nothing to worry about. Why, after all, would He ever create a world that could have too many people?

This in any case would seem to be the conclusion to be drawn from Beisner's exposition of the current state of ecological affairs. But he does not go so far as to say that we can do what we want with the environment (on our "authority"); he does allow that "the wisdom [of clear-cutting] and the safe levels [of CFC emissions] are to be determined by scientific investigation, not by appeal to Scripture" (50). Note then that Scripture, ultimately, has *nothing to say* about environmental issues; those are scientific questions. In a similar vein, Beisner notes that "the principles of Biblical stewardship... involve a free market within the moral restraints of God's law" (145).

We can conclude from this that, for Beisner, God's text, His Law, has nothing to say about the details of human authority on Earth: the environment more or less takes care of itself, since man was put on Earth to populate it, and if there are environmental problems there is always science; the "free" market really does run without the necessity of God's invisible hand (aside from some "moral restraints": the Ten Commandments, no doubt). God's Law as presented in Scripture is therefore a closed loop, a Law that is concerned primarily with the necessity of recognition of... God's Law. Without that one recognition all is disqualified as idolatry. With it, things take care of themselves. The content of Scripture boils down to the need to recognize the truth of Scripture, to take it literally, and to repeat it by citing and teaching it. As Bob Jones Jr. would say, the text explains itself, and our role as believers is nothing more than to read the text reading (and affirming) itself. If we do that, the environment and the economy, with a little help from (creationist and free market) science, will take care of themselves, virtually independent of God's Law.

One is reminded here of Kojève's "absolute knowledge," embodied by a Book that contains the Truth but that can be read at the end of history only as sheer repetition. The Book is "dead" in its definitive status, its imperviousness to change and its indifference to societal mutations—so much so that Kojève can posit, as an example of posthistorical activity, a society (Japan) that is completely ignorant of the Book. As with Kojève, literalist biblical truth is self-referential: God is nothing more than the textual embodiment of the necessity of reading and affirming God. Everything else exists in a kind of textual pocket affirmed by Scripture but in which Scripture has no sway: the details of environmental stewardship and the vicissitudes of the free market. God as the supreme instance of textual Law is above man, existing independent of him, indifferent to creation except insofar as man continues to read and affirm him and his Law. True, this fundamentally Old Testament God will ruin the environment if man stops reading and repeating, but beyond that everything else is man's concern. The ditto-head can feel comfortable with his fossil fuel lifestyle as long as he keeps "reading" his Book. While Beisner would never posit a society, like Kojève's Japanese, that is completely ignorant of Scripture yet living under its sway, his model of the faithful nevertheless entails the performance of simple reading machines, affirming God as they affirm the act of textual repetition. This is an act of reading that presupposes an absolute fidelity and accuracy that has nothing to do with the acts of interpretation and consensus building that characterize reading as it has come to be understood in the twentieth century.[9] That has nothing to do, in other words, with time. Beyond this robotic and ephemeral repetition, partitioned off from any temporally located and limited practice of reading, the world continues to evolve, indifferent to the exigencies of the text. Only the fear of an otherwise incomprehensible environmental catastrophe keeps the incessant text act functioning and people in the pews worshipping—that is, unless the faithful decide that the Apocalypse will soon be upon us, and that the destruction of the heathens—the Rapture—is past due.

Bataille: The Text, the Law, and Animals

We have already seen, in chapter 3, Bataille's rewriting of the sacred Book and its opening out through rereadings of Sade, Kojève, or the Bible, what I called his counter-Book. Now, thinking of the future, I would like to consider how Bataille might be rewritten to open out current models of the Book: the humanistic Book (ecoreligion) and the literalist Book (evangelical environmentalism).[10]

That these two models—religious ecohumanism and evangelical literalism—inadvertently rewrite Sade and Kojève allows us to continue Bataille's critique of these "masters" in a contemporary register. Sade and Kojève's avatars, in fact, seem to dominate contemporary models of religion. We can save the planet and ourselves through a human-centered religion and a Book cruelly indifferent to Man's bad faith. Or we can reject Man as a construct entirely and return to a fundamentalist religion and a literalist Book, one that sees no distinction between God, text, and Law, between the realm of the divine and the activities of everyday life, activities that are in turn strangely free in the end to ignore the Book, repeating or worshipping it in splendid isolation. The Books of humanism and literalism are ultimately (cruelly or benignly) indifferent to human life on Earth, to any attempt at rewriting excess through science and religion.

But for Bataille, the death of God is the point at the summit of creation in which God's name refers to *nothing* because God is not subordinate to human needs and demands. God is not necessarily rational, is not necessarily consistent, is not even necessarily good; God certainly does not make it a habit of rewarding or punishing Man. The old boy—or girl—is not, in other words, the projection of human demands on the universe. She or he is not universal; his or her absence is not the indication that she or he transcends everything but rather that she or he is *dead.* God's own atheism opens the system, and the human identification with the void at the summit, the empty space of God, entails a fall, a brutal movement downward from sense, from idealism, from a coherent materialism that posits a graspable, practical universe with Man as the privileged being who understands. God and Man fall together, disbelieve in themselves together. Bataille writes:

> *God's absence* is no longer closure: it is the opening of the infinite. *God's absence* is greater, it is more divine than God (I am thus no longer *I,* but the *absence of I:* I was waiting for that conjuring away and now, beyond measure, I am gay). ("L'absence de mythe," *OC,* 12: 236; italics Bataille's)

At the summit of a monotheist creation, a comprehensible, erect universe with docile matter fully under our control—the point of ultimate indifference to God—we "fall" not simply into a void, a black night, but through the openings of our bodies and of others into a profound if mortal generosity. The integrity of our selves, our bodies, our books, the Book, is rent, and we "communicate," in Bataille's sense, with that which cannot be controlled or grasped. Hence the "dualist materialism" of Bataille (Hollier 1990): at the highest point not one God but a God absent to himself; not one Man but two lovers, torn apart, doubled through their mortal wounds; not one comprehensible matter, but cursed matter, filth, virulent with its incessant energy flow, the "circular agitation" of its atoms.

Man, then, is rent through the reading, the hole, of Bataille's counter-Book. The text makes no pretense of providing a Law by which the apparent randomness and chaos of the universe, or of human society, can be subdued; the sovereign, forsaking mastery, forsaking the heterogeneous position at the summit that epitomizes and renders coherent the expenditure of society, offers not Law, but complicity:

> Sovereignty is either silent or disposed. Something is corrupt when the "sovereign" gives explanations and tries to draw inspiration from justice.
>
> Saintliness that is coming is thirsty for injustice.
>
> The one who talks about justice is himself justice;
>
> He suggests to his fellow men an upholder of the Law, a father, a guide.
>
> I could never suggest any justice.
>
> My complicit friendship: here is all my character can bring to other men.
>
> A feeling of wild celebration, licentiousness and puerile pleasures determine my relation to them. ("L'Amitié," *OC,* 6: 303; "Friendship," 2001, 12)

The sovereign is a double of God; he is the nonupholder of the Law, a nonfather, a nonguide. "I do not believe in God because I do not believe in myself," Bataille writes, and the sovereign is nothing more than the act, if one can call it that, of being "accountable to no one" ("L'Amitié," *OC,* 6: 303; "Friendship," 2001, 13). Disbelief in God is inseparable from disbelief in Man, the avid and closed self. The sovereign is not to be confused with, say, a medieval sovereign, who, while perhaps above specific Laws to which his subjects were bound, nevertheless was responsible for them as leader or feudal lord.[11] Bataille's sovereign is responsible to and for no one, radically unconditioned, *heterogeneous,* as Bataille would put it, to the requirements of any and all constructive life. One version of this sovereign is nothing more than the postautobiographical Bataille, writing.

The sovereign, like the atheistic (dead) God, is not complete, but recognizes his own and the other's incompletion: that is what links him to the other, in friendship, in eroticism. Bataille, again in "Friendship," writes:

> If the illusion of completion is not adequately rendered, in its totality and abstraction, in the representation of God, but rather evoked, more humanly, in the presence of an elegant but partially clothed woman, her animality is glimpsed again and her gaze delivers in me my incompletion. . . . It is insofar as existences appear to be perfect, complete, that they remain separate and closed in on themselves. Existences only open up through the wound of the incompletion of being in them. ("L'Amitié," *OC,* 6: 295–96; "Friendship," 2001, 6)

Note that the "rendering" of completion appears impossible; its failure leads to "evocation" rather than representation or reproduction. But what would evidently first evoke completion—"an elegant but partially clothed woman"—immediately evokes something quite different. A woman's beauty, it would seem, while first appearing to suggest an ideal perfection, leads instead, via nudity and the vision of aroused organs, to a recognition of her own "animality." This in turn, through a play of recognition, leads the narrator to recognize his own incompletion.

There is a kind of relay operating here between a beautiful woman, animality, and the subjectivity (and its loss) of the observer. Just as attainment of absolute knowledge is immediately reversed into non-knowledge, so too the recognition of the beauty and perfection of the other, her completion, leads to a recognition of her animality. The animal here is specifically the sign, the mark, of the incomplete. Hegelian recognition of the other is rewritten as the recognition of the other as radically wounded, open; the animal is the movement down from the summit, from the perfection of God, to an earlier, bestial state. More important, the animal is a being who is fully integrated into the natural order (the order of simple appropriation and expenditure of energy) rather than a being, like Man, who consciously negates the environment, given being, to create a stable and comprehensible—human—world.

In the other, Bataille would recognize a being who cannot be recognized: not a woman as partner, as interlocutor, but a woman as sexually aroused and arousing, who *stares back.* In the moment of recognition there is a nonrecognition: what is conveyed to me is not recognition of myself, my liberty, or my status as agent; it is instead my status as double of a denuded aroused woman, the woman as like me but animal, the "communication" to me of myself as animal but also the "communication" to her of myself as animal (see Guerlac 1990). Out of this "communication" there arises a doubling, a duality, that entails not recognition as such, but rather a "mutual absence."[12]

The woman and the animal: the most pure and the least pure. I am the anguished double of this woman if I am a man. It is difficult to see, if Bataille's figure of the woman were rewritten by a woman, how the woman could not be repositioned as a man, perfect, noble, majestic, or beautiful before the gaze of the woman—and torn open in turn, degraded by animality. As Shannon Winnubst (2007) has argued, "sovereign" gender roles in Bataille are in incessant flux.

Bataille's counter-Book, his sacred text, can evoke not an irreproachable Godhead but the woman as double. Mme Edwarda's nudity not only arouses but strips from the other the pretensions to wholeness, identity: seriousness of purpose, calm intellection, the authority of Law and tradition. One is reminded of Bataille's discussion of the controversy that surrounded Manet's painting *Olympia* (1863) when it was first shown. What scandalized in Olympia's nudity was its challenge to all tradition, all authority, all dignity, all eloquence tied to human institutions (aesthetic, legal, religious):

> What dominates, if we look at Olympia, is the feeling of a suppression, the exactitude *[précision]* of a charm in its purest state, that of an existence having, in a sovereign, silent manner, cut the bond which tied it to the lies that eloquence had created. (*Manet, OC,* 9: 142; 1955b, 67)

Olympia's fearless gaze, which holds us, fascinated, aroused, is the gaze of the destruction of the elevation of human culture. And yet, of course, this is accomplished through a representation in a painting, an art object of the greatest prestige. At the summit (Olympus/Olympia), at the pinnacle of the eloquence of culture, there is the troubling gaze of the nude and ultimately disrespectful woman. Her incompleteness, doubling and inciting the observer's incompleteness, is inseparable from that elevation to and fall from the monument of human and divine civilization, from eloquence, from the Book as unchanging codification of divine Law and human culture.

Bataille's "communication" between two beings torn by, yet united through, their intimate wounds is triggered by an evocation of completeness (perfection, beauty)—a mimicking of absolute knowledge—and a fall into the erotic, into death, into a non-knowledge of animality (quite different nevertheless from Kojève's animality).

For the animal, I argue, has a privileged place in Bataille's work, and his notion of it serves as a powerful response to the conceptions of nature put forward by, among others, the ecotheologians considered earlier in this chapter. Just as Bataille's Book is open to "communication," wounded, aroused, mortal, so too his ecology is inhabited by creatures engaging with humans in an intimate, material, death-bound way.

In his book *Lascaux, or The Birth of Art,* a companion to the work on Manet (both were first published in the large-format, illustrated Skira series) Bataille considers why so many Neolithic cave paintings focus on animals and why representations of the human are so sketchy, mere stick figures for the most part. His conclusion is that the paintings were not mere illustrations or decorations but were meant to play a part in sacred or magic rituals. The paintings served, in other words, in rites in which humans reaffirmed the special status of the animals they were hunting. Moreover, humans hoped, through their contact with animals, to tap into a world that they had forsaken. Bataille starts by citing Eveline Lot-Falck, the author of a book on the hunting practices of Neolithic Siberians: "The animal is in more direct contact with divinity, it is closer than man to the forces of nature, which it willingly incarnates" (*OC,* 9: 75; *Lascaux,* 1955a, 126).

In this sense, early hunters instinctively recognized the "superiority" of the animal. This superiority was due to the fact that the animal had never broken the continuity of nature, had never done precisely what Kojève argued was human: radically negate nature by working, transforming it. In this sense the painters of Lascaux, according to Bataille, were already attempting to compensate for the alienation from nature that went hand in hand with planning, conservation, and labor—with, in other words, history, language, and knowledge. They were already taking into account Kojève's argument and recognizing the status of the "human" as one of profound alienation. Bataille writes:

> It [in the cave paintings] was always a question of negating man, in so far as he worked and calculated, by working, the efficacy of his material acts; it was a question of denying man in favor of a divine and impersonal element, tied to the animal who does not reason or work. (*OC,* 9: 69–70; *Lascaux,* 1955a, 121)

The earliest art, then, was, along with ritual (the two could not in fact be easily separated), man's way of regaining what he had lost by becoming human. Labor, time, history, all this was in a sense a betrayal of nature, and it induced feelings of guilt. Bataille stresses for this reason that the affirmation of and identification with the animal is not so much "primitive" as it is more sophisticated than mere (modern) humanity: it is a phase of human development that comes *after* the merely human, since it is a recognition of the limitations and violence of constructive human acts.

> Humanity must have had the feeling of destroying a natural order by introducing the reasoned action of work; it acted as if it had to ask pardon for this calculating attitude, which gave it a true power.... From the time of the hominids, work took place, logically, following principles contrary to the supposed "primitive mentality" that was supposed to have been "prelogical." However, the conduct that is dubbed "primitive" and prelogical, which is really secondary and postlogical, magic or religious activities, only serve to translate the uneasiness and anguish *[l'angoisse]* which had seized upon men acting reasonably, conforming to the logic implied in all work. (*OC,* 9: 70; *Lascaux,* 1955a, 121)

The cave painters' identification with animals could for this reason be compared to the literally postmodern, post-Kojèvian position that Bataille worked out in *Inner Experience.* Already in prehistory humans had bumped up against the radical limitation represented by work and the time it made necessary: a time of deliberation, postponement, planning. Like absolute knowledge incessantly tipping into non-knowledge, the cave painters' humanity, their logic, through their repeated magic rites and sacred rituals moved back, identified with, and lost itself in anguish, in the continuity of the animal. Their highest, "postlogical" practice was actually a profound identification with, loss in, the alogical, the force of the universe unconcerned with the accumulation of wealth, the strategies of survival, and an arrival at some phantasmic point of completion and plenitude.

Yet Bataille too recognizes that this identification with the animals was also part of a quite practical activity. From the start, the postlogical was never simply separable from the logical. Identification with animals in their continuity, their distance from human effort, at the same time implied the carrying out of a human task since the paintings were undoubtedly part of a hunting ritual (and they depict the hunt, with wounded and dying animals). Bataille recognizes this duality, this inseparability of planned action and sovereignty.

> These [animal] figures expressed the moment in which man affirmed the greater value of the sacred *[sainteté]* that the animal must have had: the animal whose friendship he perhaps sought, thereby concealing the crude desire for sustenance that drove him on. The hypocrisy that caused him to conceal this desire had a profound sense: it was the recognition of a sovereign value. The ambiguity of these actions betrayed a major sentiment: man judged himself incapable of attaining the desired goal if he could not rise above himself. He had at least to feign equality with a power that surpassed him, which calculated nothing, never toiled, was always at play, and whose animality was not distinct. (*OC,* 9: 78; *Lascaux,* 1955a, 127–28)

This is a complex gesture, and Bataille is not by any means simply condemning the "hypocrisy" of the "postlogical" hunters and celebrants. Man cheats, but in so doing recognizes and affirms what is higher. The very act of cheating, of assuring his survival by planning and work (the coordination of the hunt), implies another activity, a sacred or magic rite, through which man conceals from himself his true motives. Or put another way, he hunts, but as a sacred ritual: he can only attain his goal—eating, surviving—by *not* attaining it, by doing something else. The hunt is the recognition and attainment of a "sovereign value," and the assurance of physical survival. Hypocrisy, perhaps, but Bataille again makes it clear that the attainment of knowledge, the goal, is never separable from non-knowledge, the sovereign state beyond work, beyond separation from the void. The postlogical cannot be simply and clearly delineated from the logical; there is a "circular agitation" in which we move from one to the other. We might even argue that survival is the unplanned aftereffect of the sacred rite; the gesture that risks the integrity of the human is also the one that assures it. Humans know which value is "higher"—but they also see, inevitably, that an unconditioned sovereign act, one in which, say, knowledge would play no part, is unthinkable, meaningless. The sovereign act, in other words, has consequences in the world, consequences whose effects are inevitably taken into account, affirmed. But the goal is logically subordinated to the unconditioned (sovereign). The hunter hunts to live so that he can practice the rite; he does not practice the rite only so that he can survive. At the moment of the kill the mere effort to survive is forgotten, even discredited. Thus through the sovereign act the hunter assures the survival of the community (the community of lovers), but that survival is justified not on practical grounds, but on "higher" grounds. Survival is the mere spin-off of the sovereign act, but could the community survive without that act? If the act of hunting were purely practical, could humans survive? Perhaps not; though Bataille does not state it, we might conclude that a radical separation from the continuity of nature, from the sacredness of animals, would condemn man to a state of logic in which he not only feels guilty, but *is* guilty. This is the state of the human as closed off (from the excess of the universe), falsely permanent, falsely self-sufficient, the ultimate abstract signified that is the *end* of all work: a living lie (of permanence, plenitude, necessity). Living that lie, the lie of sheer utility, would be impossible in the long run. Hence ecstatic/anguished religion, the sacrificial moment, the *end* now, affirms sustainable activity (the procurement of food in this case) only as a subordinate adjunct to the unconditional (impure) sacred.

Closed off, the phantasmic self would engage and serve as the guarantee of appropriation, conservation, reinvestment. All laudable activities, no doubt, but activities that in the end require a lancing, a loss so that once again, "postlogically," "the human is lost in the festival of suns and spirals," as Bataille put it in the prewar essay "Celestial Bodies":

> An avid existence, having arrived at the highest degree of growth, reaches a point of disequilibrium and it spends, suddenly, with prodigality; it loses, explosively, the increase in forces that it had amassed with difficulty. . . . The useful [thus] becomes the subordinate and the slave of loss. (*OC*, 1: 519)

One can see this movement temporally organized in a succession—buildup, then the blowout, as here—or one can see it as simultaneous, as in the case of the Lascaux hunters (sustainability-sustenance and the ritual of expenditure, together). In the latter case the movement of loss is coterminous with that of accumulation: the hunter hunts to eat, but he hunts as a ritual that celebrates the superiority, the sovereignty, of the animal. The act of hunting, in conformity with the logic of conservation and accumulation, nevertheless participates in, is inseparable from, an act that identifies the human not with mere survival but with the loss of the human in the vastness of creation. In the terms put forward in *Inner Experience,* we might say that sense, the coherent meaning of hunting for food, is both carried out, accomplished, and at the same time radically lost through its transfer to non-sense, when it becomes the ritual in which the hunter "becomes" the animal. Perhaps the hunters, like the narrator of *Inner Experience,* saw the ritual of the hunt as a kind of miming of hunting in which the activity of the human is carried out to the point where the human in its differentiation goes down to defeat. "Postlogical" hunting.

We need to read Bataille critically if we are to "use" him to think about the future—as he did with Sade. It turns out that his Book, his *Summa Atheologica,* does indeed propose a model of ecology, of human intervention in nature. This is neither the nature of the ecoreligionists, respected as a sacred domain, protected from our base consumerist instincts, nor is it that of the literalists, for whom nature is a seemingly inert and limitless warehouse because God wants it that way.

Instead, we are invited to think of a de-anthropomorphized nature quite different from that conceived by the thinkers of "deep ecology." This movement of ecological theory, first proposed by the philosopher Arne Naess, posits a world whose resources are more than just material for human use. (Heidegger's influence here as well is not hard to detect.) Naess himself is fairly commonsensical: for example, in his influential essay "'The Shallow and the Deep': Long-Range Ecology Movements, a Summary" (1995b), he makes a crucial distinction between "shallow" ecology, an environmentalism devoted to nothing more than cleaning up the planet so that "the health and affluence of people in the developed countries" will be preserved, and "deep ecology," which seeks to preserve the earth not for human convenience but because it is of value in and of itself. This entails a process of de-anthropomorphization: Naess's goals are, however, fairly modest. He recognizes that "any realistic praxis necessitates some killing, exploitation and suppression" (151) and suggests lifestyle changes such as "use of simple means," "absence or low degree of 'novophilia—the love of what is new merely because it is new,'" and refusal to "use life forms merely as means" (1995a, 260–61). In other words, the "deep ecology" imperative is one of simplifying and making human life more reverential toward all biologic systems (and in this it resembles ecoreligion), but it is hardly one of the extirpation of human life and the cult of pristine, impossibly human-free ecosystems.

The evil twin of deep ecology, at least in the minds of journalists, is the Earth First! movement, which would use nonviolent and even (targeted) violent actions to put a halt to development, logging, etc. Here de-anthropomorphization manifests itself on a practical level: Earth First!ers see humankind as simply another animal population, subject to the same constraints as any other. From this pseudoscientific and highly contested vantage point, the human population "bloom" must be controlled, and ecosystems must be preserved from the ravages of a species run amok. A human life is not intrinsically more valuable than, say, the life of a bear.[13] This variety of antihumanism signals more than a lifestyle change: it involves the total or partial cleansing of certain ecosystems of their human population. The ideal, in some cases, is the return to a pristine nature untouched by man.

Even on the "radical" end of ecology, one can see that critics have a hard time negotiating between the demands of human life (deep ecology as lifestyle) and the demands of a (supposed) "saved" ecology rid of the sin of human existence. Yet one has the sense that even the Earth First! position implies a certain anthropocentrism: are the perfect, human-free forests an object of desire precisely because they offer the perfect consumer fantasy of nature? To what extent is nature untouched by human hands nothing but an eminently human fantasy, like any consumer object whose beauty resides in the fact that it has not and never can be used? (Collectors value most highly objects—coins, cars, whatever—in "mint" condition.)

A Bataillean ecology, on the other hand, would put forward another version of the natural, and of the animal, in the context of the death of Man (which should not be confused with simple de-anthropomorphization, the phantasmic evacuation of the human). In other words, it could help us rethink deep ecology by reformulating the critique of the role of the human in nature. The connection with animality in Bataille is sacrificial: it recognizes the animal, it values it, neither as mere matter to be consumed, nor as an element of a good lifestyle choice, nor as imaginary (untouchable) fetish object. The animal in Bataille is a creature that looks back, that stares into one's eyes, but the recognition is not one of mutual liberties, of human qualities tied to constructive projects, no matter how sensuous.[14]

The look between animal and human, the profound bond, is a violent one: it is the movement of capture, killing, eating. The animal is sacred, sovereign: its existence for the human is one that ignores the world of duty and discipline. By killing the animal, we partake of, revel in, that world, the emptiness of the universal of the animal's creation. We are sovereign through the animal, just as we are sovereign through the writing, reading, and rewriting-rereading of the counter-Book: through, in other words, an impossible recognition (of animal continuity, of non-knowledge).

The sacrifice of animals as nonhumanized beings—as creatures outside the easy domination and use by humans—implies a model of ecology that is perhaps as close to that of the Lascaux hunters as it is to that of modern workers and intellectuals. Obscene animals are to be (impossibly) recognized, sacrificed, as wild, like the aurochs or elk of Neolithic caves Bataille explored. They, unlike us, are still in profound communication with an order not of avaricious concentration but of open energy flows, an order where densities of energy break out rather than lend themselves to use. We join the animals, but not by breeding them on industrial farms and quantifying the weight of their meat. A sacrificial relation between humans, animals, plants—the ecosystem—is one of the affirmation of proliferation and of the recognition of the relatively minor position of humanity, finally, in the concentration and expenditure of the energy of the universe. Humans may be the most intense storers and squanderers of energy on the planet, but their squandering only puts them in intimate contact with other life-forms, also engaged in storing and squandering. The toppling of Man is the recognition that we are in the next energy regime—the post–fossil fuel regime of scarcity and agony—together, even in death.

A Bataillean ecoreligion, if we can call it that, would perhaps be a version of deep ecology (if that implies a critique of an all-embracing anthropocentrism), but one that recognizes the natural as a realm not simply autonomous from the human, nor simply subordinate to human desires and "needs," but instead as one that directly challenges the human in its deepest, most characteristic practices. Nature is not simply autonomous; it is the intimate world in which the human trespasses, a world that challenges the human, opens it out in anguish and ecstasy. It is more than autonomous; it is radically foreign, a world of continuity in which the animal lives and spends but does not work and plan. Ecoreligion in this sense is a field of sacrificial expenditure, not a doctrine of the (now) respectful managing of raw materials under the aegis of Man or God: the human enters into this world neither to exploit nor to appreciate from a distance (nor to feel that the world is not coming to an end), but to die as human, to sacrifice him- or herself in a transgression of human limits. The human community's physical survival (through sacrificial consumption) in this model is the fundamentally unplanned aftereffect of a sacred "communication" with the animal. Man's death is thus a "natural" event, the recognition of the animality of the human, the ultimate defeat of objectification and planning, the loss of Man as "cause, principle, end" in the labyrinth of sacrificial (bodily) expenditure. I would argue that it is only in this sense that religious debates concerning ecology can be posed: from the perspective of the death of God and of what accompanies it, of what is indeed inseparable from it, the death of Man in sacrificial communion.

Beyond this, in a contemporary setting, one enters into communication with the sacred animal, with base matter, with sovereignty, as reader and writer of the counter-Book. Bataille's *Summa Atheologica,* rewritten as a postecology, continues to spin off rewritings (like *this* book) that double, and parody, the Books (Sade, Kojève, the Bible) afflicting the twenty-first century.

7 An Unknowable Future?

Expenditure in a Time of Depletion

The City, the Car, and Alternative Modes of Transport

Bataille, as we have seen, tends to think of intellectual or theological constructs in the framework of spatial organization and structure; hierarchy implies not just a ladder of concepts, but a physical ascension: the elevation of a building, the erection of a statue at the center of a plaza or city. A coherent concept is never a simple abstraction; for Bataille, the height associated with a god or leader is quite literally the rising above of a representation or a human body. The eagle, symbol of eternal empire, is carried above the advancing troops; the fascist dictator stands above the throng and screams his orders. Even coherent ideas are conceived vertically and visually: mathematical ideas are a "frock coat"—thus the Bataillean critique of architecture analyzed by Denis Hollier. In addition to architecture, however, Bataille also sets his sights on any physical thing that would seem to guard, at its summit, some point of absolute meaning or value that justifies and motivates all "profane" activity that somehow takes place below it. The elevated and elevating object, the *thing,* in its very vertical layout indicates the way to go: everything one does points to something else, a value, a meaning, a God, that is elsewhere, above and beyond. In that sense the lowliest practical object has the same structure as the greatest work of architecture because it points or leads to something else. If I use a hammer to carry out a productive task, the hammer, in its form, already shows that it is for something else: it is to be used to make something, which itself will be useful (a house, a piece of furniture). All these things point to some higher goal or meaning that justifies them and the labor of which they are a part. A church or statue is no different in the sense that each points—directly now—to a higher meaning that justifies its presence, and ours. A church or statue, however, embodies this pointing more directly, more visually: each literally points upward, into the blue (of noon?), to the sun, out of which meaning is generated—and lost. When the emptiness, sovereignty, formlessness of this highest point is *recognized,* it collapses, dissolves, and the seemingly stable conceptual and physical world erected in its name goes with it. For Bataille the entire edifice of civilization amounts to an enormous misreading of the unconditioned: if the latter really is sovereign, it justifies nothing, motivates nothing, serves as the raison d'être of nothing. It is heterogeneous and base. At best it is a useful lie, recognized for what it is, useful only as long as its usefulness is subordinated to what it "really" is: the impossibility, the perversion, of being itself. The radical recognition of sovereignty—which is not an intellectual operation, but rather the collapse of the intellectual in laughter—for Bataille is the religious "experience."

For Bataille, the theocentric city is a lateral construct that is nevertheless inseparable from the hierarchical, fixed form of the monument. The city could be said to be, on the horizontal plane, the adjunct of the verticality of the monument. The city is the base, the obligatory spatial organization that underlies that which rises above. The city draws in people and goods from the boondocks; it concentrates them in central places where they are bought, sold, stockpiled, and sacrificed.

As we have seen (in chapter 4), the modern city for Bataille is a problematic space because it transforms, translates, this centrifugal movement to a centripetal one. The markers of the elevated, the timeless, the holy, have become so abstract, so ideal, volatilized, that they are on the point of toppling. They no longer fulfill their role of maintaining and strengthening the profane order that supports them: through their absence, they deprive the secular order of the coherence necessary to its sense of purpose. The markers that stand in for the gods, for God, reveal themselves to be what they always were: unseen, not worthy of being seen, meaningless, radically heterogeneous to the profane order but no longer tied to a transcendent realm capable of justifying activity. In the indifferent atheism of the modern world, God, the sun, or whatever, has revealed itself as the dead node around which activity—indifferent motorized traffic—turns. It is just a matter of recognizing the emptiness of the ultimate signified—and the whole structure will crash. It is as if by intensifying, speeding up the movement of a theocentric, logical, and legal world, by making the seemingly isolated "particles" spin around the center at an ever faster rate, modernity has revealed the emptiness of the universal signified. The universal, the sovereign, was empty all along, base in its heterogeneity, a capstone meaning nothing, leading to nothing else, providing nothing, but it had always been taken, at least in the Christian tradition, as a plenitude, a holy, pristine guarantee of permanence and happiness. Now it is what it is, when we have really witnessed it, really circled it, moving as fast as we can go (in our cars): nothing, that which leads nowhere else, means nothing else. How much longer then can the cars blindly rush? Why circle what does not exist?

We have followed the movement of the elevated sacred, mimed it in our cars, parodied it, gone all the way with it, and it has, or is about to, collapse. The world of conservation and concentration is about to blow. The madman—Bataille himself, the narrator that mimes "Bataille"—stands in the public square, at the very center, and inaugurates the next phase of the drama: the fall, the "tumble" *(dégringolade)* back down and out—to the base of the monument, to the most remote locale of the most outlying region. Moving up and in, miming the arrival at perfect knowledge and divinity, infinite resource stockpiling and disposal, is inseparable from a crash: death, infinite wandering. Another era, of violence, deicide, sacrificial (dis)aggregation, is about to begin: the era of the recognition of the limit, of the deleterious movement of time.

This central and elevated point is outside the "community it governs": as Jean-Joseph Goux puts it, it "legislates as an exception" (1990b, 31). The general equivalent, be it gold, phallus, monarch, father—or the Law of the elevated and selfish subjectivity—is a mediating term, the equivalent of everything, specifically signifying nothing because it transcends all, does nothing; it is different from all in that it is the space of identity. Gold overflows; it can "mean" anything (wheat, iron, slaves, books), and yet in its universal meaning it is excessive, beyond mere equivalence because exceptional, exemplary. Goux goes on:

> Commodities are universally evaluated only through the detour of *specie*—that is, through signs, masks, representations. The function of common denomination in all registers is therefore linked to the sacred position of the general equivalent. The general equivalent is the site of a law, the place from which the name is imposed. (38)

The "site of law": the collapse of the privileged, elevated space is also the moment in which law is reversed. The festival of crime, the eruption of the left-hand sacred, is the point at which the symbols of divinity, paternity, and wealth seize up, shatter, scatter.

Public, organized space, then, can be formulated as commerce, translation, and the organization of desire in the law of the father. But, as Goux reminds us, this organization is itself a play of masks, of empty signs: acceding to them, we recognize their emptiness, their fragility. This would seem, in Bataille, to be the role of the automobile: the traffic circling the Place de la Concorde, the taxi in which Mme Edwarda takes on the driver, are vehicles for a rapid movement that indicates the imminence of the death of God, the atheism of God. Mme Edwarda, as God, steps into the taxi, stops the movement, releases time, but in that sudden paralysis her madness reveals the emptiness of the values, the divinities—the specie, the masks—*that were already empty.*

The car, then, for Bataille, is the ultimate secular mover, the device whose speed illuminates the fall of the statue, the tumble of the elevated divinity—*in general indifference.* It is worth considering, however, in more detail how the car brings about this fall. Quantified and quantifiable technology, epitomized by the automobile, first presents itself as the empty signifier par excellence, replacing gold as the standard measure and excess of the economy. The car resituates the city to its interior: the public square, the central urban space, is no longer that which the car circles around—as it was in Bataille's day—so much as it is the point, the particle, reproduced to infinity in all the cars of the world. The square and the self—that of the driver and his or her passenger(s)—are coterminous: where the car is, there is pure space. As Paul Virilio has pointed out, with speed comes a paradoxical exhaustion of time: the faster one goes, the less the past and future count. "Going quickly" is precisely the suppression of waiting and the passage of time:

> Man will be delivered from the apprehension of the future which will no longer have a reason to exist, since everything is already there, here and now, both present and disappeared, in the instantaneous apocalypse of messages, of images, in the happy pleasantry of the end of the world! (Virilio 1993, 123)

In high-speed travel we lose the sensorial experience of movement: instead there is the merely visual passage of the surroundings, "the projection of a film onto a big screen" (116). Moreover, the movement of this landscape freezes, disappears in the "distance of altitude."

The car is the device par excellence that is capable of reproducing a subjectivity bound in the pure vision of the now. Speed is the revelation of an unchanging presence, an object presented to a subjectivity that, encased in metal, becomes an object to other subjectivities. The timelessness of this mechanically reproduced subject is the fulfillment of the promise of the obelisk: it is the signification of the ultimate fulfillment of a godhead, a pure meaning, not as elevated divinity but as mechanically reproduced identity, sheer indifference.

The automobile—the self-movement of technology—becomes, by metonymy, the empty signifier at the summit of fossil fueled modernity. It is speed, the mastering of time, the empty and necessary eternity—comfort—of the encased, packaged, identically emptied and reproduced subjectivity. Reality is an always but never changing image on the (wind)screen: the obelisk glanced at, indifferently. Everything is mediated through the automobile, everything translates into everything else, but the car itself is empty, the excess that in itself means nothing other than an empty now and an empty space from which there is an empty never/ever changing vision of space. And then the car, opened out by the death of God, falls, and Man ("freedom") falls with it.

As the ultimate common denominator, the car brings together, in the isolation of vapid subjectivity, social classes and identities. All are one on the freeway, mixing while not mixing, moving around the empty circuit of gutted urban space. All is mediated through the automobile: everyone derives the meaning of their lives through it: as a status marker, as a simulacrum of the freedom of movement and consumption (David Brooks's utopia), as the timelessness of a religion shared by all. Virtually everyone in American society works as hard as they do to pay for their cars: it is their major investment, the acquisition that justifies and represents human labor ("drive to work, work to drive"). And, as common denominator, it is the transcendence of labor, of meaning: sitting in traffic jams, the driver does nothing, just sits, and in this way lavishly neutralizes the labor devoted to purchasing the vehicle.

Finally, pure space leads to a triumph not only over time but over the body, and bodies as well. In the car we do not need a body, we have no thought for its energy flows and expenditures. Cursed flesh is miraculously transformed into an idea. The body's energy is stored as immense amounts of fat, it can barely move on its own, barely breathe; fewer and fewer people notice. And the bodies of others are also derealized; we see no one else's body in their car, just indifferent heads, and as we zip past ghettoes made possible only through the judicious construction of freeways, the hypertensive bodies of people of other colors are happily ignored as well.

The car is the temple, the ultimate monument, of the empty self, the all-powerful self in its pure objectivity/subjectivity. Like gold, like the father, the phallus, it brings all together, signifies all, is invisible in its materiality, and refers to a transcendent but empty signified.[1]

When we gaze at this speed, really lose ourselves in it, the great abstraction to which the car is devoted, its god, the human self, collapses: it is nothing but a pure now, a pure vision of the different, which is always the same. The self "sees" only urban space, the same interchanges, the same off-ramps, the same blur of buildings passing in the night.[2]

But at the heart of this empty plenitude, this universal in which all things are recognized but only as subsidiary, indistinguishable moments, there is a limit, a cut, a finitude. It is the temporality, the death, at the heart of empty ideality, automobility, self-erected as law. Mme Edwarda puts the brakes to it in her impossible and carnal knowledge.

The car runs on fuel. It burns it, consumes it, and it can only move if the fuel is burnable, if the fuel is finite. The car's universality, and the self on which it depends, are a function of an energy that can be expended and that will soon be gone: the car's fall. At the moment of the recognition of the finitude of fuel, the space of the car opens out to another space, the space of another expenditure: that of the walker, dancer, or cyclist in the city; the flaneur, the voyeur, the exhibitionist. The one who lurks under the arch. The finitude of one energy regime, one model of expenditure, opens the way to another. As we saw in the *Accursed Share,* it is the limit that inscribes the cut of excessive energy. One limit to energy, based in a fundamental scarcity, entails another burn-off, another non-knowledge of excess, another mode of ecstatic or dreadful transport, in short, another, this time, cursed energy.

The autonomist self is inseparable from, it indeed is, the city as sketched out by Michel de Certeau in his essay "Walking in the City" (1980). There, de Certeau notes three aspects of the modern city, as it is "instituted by the utopian and urbanist discourse":

> 1. The production of a proper space: rational organization must therefore repress *[refouler]* all the physical, mental or political contamination *[pollutions]* that compromise it;

> 2. The substitution of a non-time, or of a synchronic system, to the ungraspable and stubborn resistance of tradition: unequivocal scientific strategies, made possible by a flattening of all givens, must replace the tactics of users who trifle with *[rusent avec]* "occasions" and who, through these trap-events, breaks in visibility, reintroduce everywhere opacity in history;
>
> 3. Finally, the creation of a universal and anonymous subject who is the city itself: like its political model, the State of Hobbes, it is possible to attribute to it little by little all the functions and predicates up to that point disseminated and applied to the many real subjects, groups, associations, individuals. (176; my translation)

It is not far from this city to the pure moment of the city that is the automobile: the car that produces a homogenized space, viewed and navigated exclusively from the inside; the pure present of the self at any and all (identical) points on the map; and, finally, the empty universal self, in motion but perfectly still, as product and end of the automobile, and end of history.

Against this high modern, totalizing city, de Certeau sees another practice, characterized by that of the walker, "proliferating illegitimacy, developed and insinuated in the networks of surveillance, plotting according to unreadable but stable tactics" (178).

There is, of course, a whole industry of explicators who attempt to imagine what these "tactics" of resistance to the totalizing city might be.[3] I would stress only two points. First, tactics for de Certeau are in no way simply against, or oppositional, to the city. They are instead "insinuated in the networks of surveillance," they operate within it, they are generated out of it, function in complicity with it, while they undermine it, parody it, manifest its nullity. They affirm the knowledge of the city, then recognize, moving along but against it, a non-knowledge, a parodic emptying of its impossible totality. Second, de Certeau's approach to the city, and to tactics within it, is somewhat limited by the fact that he does not consider the energy regimes that underlie both "strategies" (the universalizing city) and "tactics" (the "illegitimacy" within and against it). Like so many other critics in the Western tradition, and like virtually all thinkers of "everyday life," de Certeau posits an oppressive but utopian city without considering what energy will keep it running. Likewise, he ignores the energy in its various modes expended in tactics elaborated against it.

The universal city, one of whose greatest moments would no doubt be the *ville radieuse* of Le Corbusier, is dependent on cheap fossil fuel inputs, on the official segregation of social spaces, and on the universalized movement of the car.[4] A city with no street life whatsoever depends on the rapid movement of its idealized, derealized citizens through programmed routes determined by experts in traffic safety. Hazards, chance encounters, moments of whimsy, friendship, surreal madness, all are reduced to zero: protected from chance, the motorist is able to move along quickly, never experiencing anything other than the now and the here. This is the beauty of de Certeau's analysis, though he doesn't seem to recognize it: by positing the walker against this ideal city, he has struck upon a figure who *consumes energy differently, who spends it gloriously.* No doubt de Certeau, when he proposed the walker, was thinking of the flaneur in Baudelaire or in Benjamin. But from an early twenty-first-century perspective, there is more to this figure: he or she is moving physically, is *out of a car.* It is not just that the walker's movements are "under the radar," microscopic, rhizomatic, and therefore unpredictable, subversive, particular, peculiar. They are that, to be sure, but they are also the practice of a different kind of expenditure of energy; they are of a different energy regime.

To burn energy with one's body is grossly inefficient if one has a car at one's disposal.[5] If gas is cheap, as it always has been, and (from the perspective of the official energy experts) evidently always will be, it is inefficient to walk. You needlessly expend time, you incur physical discomfort, you are distracted by inessential things. Movement is choppy, disarticulated; you are constantly reminded of the passage of time and the finitude of your own body: death. Unfortunate surprises suddenly arise. The world is full of base matter, matter coursing with uncontrollable energy: you are confronted with disgusting smells, the vision of dirt, of rotting things in gutters. You are needlessly spending bodily energy, and time, perilously in contact with matter that could just as easily be entirely separated from the movement of a pure awareness, a pure present. Your "glad rags" get sweaty, limp, and you risk somehow coming down in the world. People might think that you can't afford to drive.

Thus more is at stake than simple strategies of resistance and complicity. The walker is using energy in a way that expends the easy certainties and the enforced legal parameters of the autonomist, "strategic" city. By walking or cycling—another way of confronting the city through the sacrificial expenditure of corporeal energy—you are passing through the car, through the logic of the car, on the way to an a-logic of energy consumption: post-sustainable transport in a spectacular waste of body energy.[6] The autonomist self has revealed its void: dependent on the car, that empty signifier,[7] the self justifies and generates a vast, coherent system of urban organization and energy consumption, a flat universe of blank walls and identical off-ramps, an absolute knowledge of pedestrian crossings and rights of way. But the self at the peak of the system is literally nothing: a simple now, an awareness, a vision, of a freeway guardrail. This self is ever changing, completely volatile, "free," but always the same particle: it can lead nowhere beyond itself, mean nothing other than itself. What is more, the self is the awareness of the gas gauge on the dashboard, which can, often does, and most certainly will read "empty."

At the height of the autonomist regime, the self is pitched into the finitude of energy depletion: walking, the spending of energy in and of the body in transports of ecstasy and dread, is the moment of temporality and mortality, the sense of the human in non-sense. The empty self is torn from its ideality; it is pure separation: "Man," enshrined in two tons of metal, is about to emerge, to fall, violently "communicating" with the death of God. As in *Mme Edwarda,* the dead God is about to get into the back seat. Every spark of combustion, the burning of every drop of gas, announces a radical finitude at the heart of seemingly endless, quantified waste.

Eroticized Recycling and Bicycling

If we uncouple the "tendency to expend" characterizing humanity from the simple consumption of huge amounts of fossil fuel–based energy—if, in other words, we posit a "good" duality in contradistinction to the current regime of the "bad"—we then can continue to affirm excess, but excess, the destruction of the *thing,* as a movement of intimacy. From the (current) "bad" duality of the automatic production of excess as a mode of utility (the gas guzzler and the "freedom" it proffers are "necessary," "useful," etc.) we pass to a "good" duality: a possible utility—the survival of the species—as an aftereffect of glorious loss.[8] Energy now will be wasted on an intimate level, that of the human body. The expenditure analyzed by Bataille, in the wake of Sade, is always on the level of corporeality: the arousal of sexual organs, the movement of muscles, the distortions of words spewing from mouths. And, we could add, using de Certeau's terminology, the expenditure of the walker/cyclist is the tactical alternative to the strategic law imposed by social and city planners, developers, disciples of autonomist Man: the vast arrayed forces of modernism in its era of imminent dissolution.[9]

There is virtually no point any more in trying to work out a critique of modernity: depletion does it for us, relentlessly, derisively, definitively. Perhaps the knowledge modernity has provided, both technical and theoretical, has been necessary; in this case the fossil fuel regime inseparable from modernity has been a necessary, if ephemeral, stage of human development. But the fall, the die-off, looms. The larger problem (entailing a task never fully undertaken by Bataille) is to think a "good" duality, the postmodern affirmation of sheer expenditure through dread and the recognition of limits (interdiction, the mortality of reference) on the scale of human muscle power and the finitude of the body. A return to the past? Not really, since the imminent depletion of fossil fuel resources will push us in that direction anyway: muscle power, body power, will be a, if not the, major component in the energy mix of the future.[10] But certainly what is imperative is an awareness that any economy *not* based on the profligate waste of resources (commonly called a "sustainable" economy) must recognize and affirm the tendency to expend, indeed be based on it. And inseparable from that tendency, as we know, are the passions, as Bataille would call them: glory, but also delirium, madness, sexual obsession. Or, perhaps closer to home, a word rarely if ever used by Bataille: freedom. Not the freedom to consume, the waste of fossil fuel inputs, but the freedom of the instant, from the task, freedom disengaged from the linkage of pleasure to a long-term, ever-receding, and largely unjustified goal. An "intimate" freedom—but not the freedom of prestige, rank, not the freedom of Man in and as security.

"Expenditure without return" is a floating concept, defined in opposition to the restrained economy whose possibility it opens but which it defies. As an end not leading outside itself, it could be anything; but what is most important is that with it there is a movement of "communication," of the breaking of the narrow limits of the (ultimately illusory) self-interested individual, and no doubt as well some form of personal or collective *transport,* enthusiasm. This concern with a *mouvement hors de soi* can no doubt be traced to Sade, but it also derives from the French sociological tradition of Durkheim, where collective enthusiasm was seen to animate public life and give personal life a larger meaning.[11] As Bataille puts it in *L'économie à la mesure de l'univers (Economy on the Scale of the Universe):*

> "You are only, and you must know it, an explosion of energy. You can't change it. All these human works around you are only an overflow of vital energy. . . . You can't deny it: the desire is in you, it's intense; you could never separate it from mankind. Essentially, the human being has the responsibility here *[a la charge ici]* to spend, in glory, what is accumulated on the earth, what is scattered by the sun. Essentially, he's a laugher, a dancer, a giver of festivals." This is clearly the only serious language. (*OC,* 7: 15–16)

Bataille's future, derived from Durkheim as well as Sade, entails a community united through common enthusiasm, effervescence, and in this sense there is some "good" glory—it is not a term that should be associated exclusively with rank or prestige. Certainly the Durkheimian model, much more orthodox and (French) Republican, favored an egalitarianism that would prevent, through its collective enthusiasm, the appearance of major social inequality. Bataille's community would continue that tradition while arguing for a "communication" much more radical in that it puts in question stable human individuality and the subordination to it of all "resources." On this score, at least, it is a radical Durkheimianism: the fusion envisaged is so complete that the very boundaries of the individual, not only of his or her personal interests but of the body as well, are ruptured in a community that would communicate through "sexual wounds." De Certeau brings to any reading of Durkheim an awareness that the effervescence of a group, its potential for "communication," is not so much a mass phenomenon, an event of social conformity and acceptance, but a "tactics" not only of resistance but of intimate burn-off and of an ecstatic movement "out of oneself." If we are to think a "communication" in the post–fossil fuel era, it will be one of local incidents, ruptures, physical feints, evasions, and expulsions (of matter, of energy, of enthusiasm, of desire)—not one of mass or collective events that only involve a resurrection of a "higher" goal or justification and a concomitant subordination of expenditure.

Yet there is nothing that is *inherently* excessive. Because waste can very easily contribute to a sense of rank, or can be subsumed as necessary investment/consumption, no empirical verification could ever take place. Heterogeneous matter—or energy—eludes the scientific gaze without being "subjective." This is the paradox of Bataille's project: the very empiricism we would like to *guarantee* a "self-consciousness" and a *pure dépense* is itself a function of a closed economy of utility and conservation (the study of a stable object for the benefit and progress of mankind, etc.). Expenditure, *dépense,* intimacy (the terms are always sliding; they are inherently unstable, for good reason) are instead functions of difference, of the inassimilable, but also, as we have seen on a number of occasions, of ethical judgment. It is a Bataillean ethics that valorizes the Marshall Plan over nuclear war and that determines that one is linked to sacrifice in all its forms, whereas the other is not. In the same way we can propose an ethics of bodily, "tactical" effort and loss. We can go so far as to say that expenditure is the determination of the social and energetic element that does not lead outside itself to some higher good or utility. Paradoxically this determination itself is ethical, because an insubordinate expenditure is an affirmation of a certain version of the posthuman as aftereffect, beyond the closed economy of the personal and beyond the social as guarantor of the personal. But such a determination does not depend on an "in-itself," on a definitive set of classifications, on a taxonomy that will guarantee the status of a certain act or of a certain politics.

Expenditure, then, plays *against*—and not through *ressentiment,* but through a difference with and a recognition and transgression of the limits of the closed economy of utility and the cult of personal satisfaction, of the personal *tout court.* If we return to our model of hyperwaste, with which we are so familiar (to the point of its invisibility), we can say that loss can be framed as inefficiency in relation to the apparent efficiency and universality of the commercial (fossil) fuels regime and the automobile that serves as its ultimate metonym. Thus, to put it simply, walking or cycling is a gross waste of time and effort when one could just drive. The expenditure of bodily energy is tied to an immediate pleasure, a *jouissance,* of spending in relation to the great closed ("global") economy of the world. Of course the "closed" economy is based on waste as well, but the cyclist *knows* her waste, revels in it, and revels in all the things she defies (and defiles) in the current economic conjuncture: not only fossil fuel use, but the logic of obesity, the regime of spectator sports (only hyperconsuming athletes are allowed physical exertion), the segregation of society by physical space and social class, the degradation of the environment in support of the production, use, and disposal of cars, and the economy of "growth" dependent on the use of ever greater quantities of depletable resources.[12] This difference with the closed, global economy, subordinated to a universal autonomist Man in the ideality of virtual time (the pure now)—this affirmation of anguish, physical pleasure/agony and "self-consciousness" in a Bataillean sense is what we might call one version of a contemporary affirmation of the general economy.

Walking and cycling year-round, if judged by contemporary standards of comfort and well-being, are a ridiculous waste of time and effort; they condemn one to a harrowing descent into "discomfort."[13] Arriving sweaty at one's job at the Department of the Treasury, after having cycled sixteen miles from Bethesda, Maryland, is the indication of a grossly inefficient expenditure of time and effort that would be better invested in tending to the details of the American economy. The worker who does this sort of thing is participating in *another* economy at the moment he or she works for the larger, inanimate fuel-fed economy headquartered (one of its heads, in any case), in Washington, D.C. The expenditure of personal energy is nevertheless tied to an immediate pleasure, a *jouissance,* of spending set against the great closed ("global") economy of the world. The cyclist's body, from the perspective of triumphant autonomist culture, is little more than an open wound, screaming for a rich energy input of fossil fuel and exposed to the contempt or aggression of the world. It is only if we see the renunciation or necessary abandonment of the car and the affirmation of muscles, in and as an economy of difference and knowledge—impossible knowledge—that this act can be put in perspective. Bodily movement as transportation, display, dance, exhaustion, passion, "communication," *all together,* in a labyrinthine urban space made dense and polysemic by the different sensory modalities of ecstatic expenditure[14]—all this entails the reinscription of "freedom," its reassignment from the sociotechnical frame[15] previously associated with the regime of hyperconsumption, social standing, and fuel depletion.

Our first questions, which are inseparable from an ethical approach to economy, will be: How and what do we expend? What model of expenditure will condition the practice of maintaining or modifying a carrying capacity? And we will inevitably think within a horizon, a series of limits, as Bataille himself does when he elaborates the ethics of expenditure; no thought can be elaborated in and as a realm of limitless, sheer waste (or sheer gift-giving). But from within, and transgressing, the ethics of limits, we have no choice but (miming Bataille) to elaborate a theory of excess in an era of radical shortage,[16] a practice of human-powered velocity in an era of gas lines, a theory of glory in and against an epoch of seemingly relentless constraint. Against the decaying urban space of experts—space that will no longer seem so expertly configured in the absence of cheap energy inputs—and the divinized space of the motorist on his or her lofty perch, the universalized SUV that doubles as the summit of the social and conceptual pyramid—another time emerges, one of the parodic recycling of castoff goods, of skills made worthless in a growth economy,[17] of lives made redundant in a society of mandatory, rigorously quantified inanimate waste. A space of the communication of wounds, a void into which the subject falls—from the eternal but evanescent peak at the center of the traffic circle to the emptiness in the opening of the Porte St. Denis. "Good" duality, in effect: the duality of an ecology of "religion" that entails not a law of the perfect human, but of the aroused animal, death-bound, confronting the other not as a pure freedom, an identical-different human face, but as an exhausted, mad, and elusive wound. The "limits to growth" are affirmed as the instance of a mortality before which the fossil fuel reserve stands (before falling): they are transgressed in the night of "transport" and "communication."

The future, then, is not "small," or "simple"; it is not an era of "lowered expectations." It is, rather, an era of *base* expectations, a swerve[18] through and against a simple movement up (God) and down (simplicity). It is figured by a regime of eroticized recycling and bicycling, the practicality of ecological awareness intensified by ecstatic and anguished non-knowledge. In his novel *Story of the Eye,* Bataille presents what can be read as a figure of this swerve:

> We soon found our bicycles and could soon offer one another the irritating and theoretically unclean sight of a naked though shod body on a machine. . . . A leather seat clung to Simone's bare cunt, which was inevitably jerked by the legs pumping up and down on the spinning pedals. Furthermore, the rear wheel vanished indefinitely to my eyes, not only in the bicycle fork but virtually in the crevice of the cyclist's naked ass: the rapid whirling of the dusty tire was also directly comparable to both the thirst in my throat and my erection, which ultimately had to plunge into the depths of the cunt sticking to the bicycle seat. . . . Yet I felt I could see her eyes, aglow in the darkness, peer back constantly, no matter how fatigued, at this breaking point of my body,

> and I realized she was jerking off more and more vehemently on the seat, which was pincered between her buttocks.... Like myself, she had not yet drained the tempest evoked by the shamelessness of her cunt, and at times she let out husky moans; she was literally torn away by joy, and her nude body was hurled upon an embankment with an awful scraping of steel on the pebbles and a piercing shriek. (*OC,* 1: 33–34; *Story of the Eye,* 1977, 38–39)

Simone, Bataille's heroine, swerves, and it is the movement, in the fall of rider and bike, of the human-powered machine as focal point of erotic energy. The machine is not only the device by which circular motion (feet on pedals) becomes useful linear motion; it is the doubling of this doubled movement with the conjunction, the copula, of erotic movement, of uncontainable and fundamentally unusable energy, with the cursed movement of a machine-human in space. The wheel disappears in the aroused cleft of bike and ass. Erotic transport and surface transport, cycling and re-cycling, energy as practical work (the movement of the bike) and as useless frenzy (the movement of the ass), coincide. Indeed, energy put to use, the bike's movement, is swallowed up by the energy of the pointless, eroticized thrusts of body parts, then re-cycled out again. The bike, an evidently practical machine, is caught up in this circular agitation, and is (momentarily) defeated in its practicality,[19] crashing and scraping, while retaining its (now erotogenic) function as muscle-powered vehicle.[20] It ends in the circulation of parody, in the contamination of language and representation by the parodic doubling of human and human-powered machine, the miming of sense in non-sense. Re-cycling as eroticized parody, in the demise of a definitive and purposive meaning (the demise of machine as tool of the human, and of the human as tool-using animal). As Bataille wrote in "The Solar Anus," "It is clear that the world is purely parodic, in other words, that each thing seen is the parody of another, or is the same thing in a deceptive form" (*OC,* 1: 81; *VE,* 5). Each thing, in any case, is imbued with pointless, erotic, sacrificial promise: the intimate world.

Globalization, Depletion, and the Future

One might respond to all this by remarking that the universal city proposed by Le Corbusier or criticized by de Certeau no longer exists. The cars in Bataille's version of the Place de la Concorde still circle, no doubt, but the emptiness of urban space, the fall of statues and towers, no longer makes much difference either way. The city, after all, is no longer real, but virtual. By this we mean that space is no longer a phenomenological given, an actual location where one finds oneself—even in a speeding car. The city as locus of movement and commerce is now everywhere and nowhere. It is the locus of Empire.

This term, as used by Michael Hardt and Antonio Negri, indicates a movement of signs and capital that is quite different from the old imperial space, which had its city, its center, and its outlying realm. Empire instead goes beyond nation-states, beyond the inner-outer distinction so necessary to any definition of a city located in real space—be it a city of squares and churches, or a city of speed and strip malls. Empire as universal city is the space of a world market, which has no outside: "the entire globe is its domain" (Hardt and Negri 2000, 190).

All is movement, contact, on a global scale. Goods and services cross space in "real time," or the closest possible simulation of it. There is no longer any private, as opposed to public, space: all space has been privatized, but privatized in such a way that it is omnipresent, neither inside nor outside. Without public space, "The place of modern liberal politics has disappeared, and thus from this perspective our postmodern and imperial society is characterized by a deficit of the political" (Hardt and Negri 2000, 188). Terms such as inner and outer, natural and artificial, real event and simulated spectacle no longer stand in opposition; working as a giant deconstructor, Empire disarticulates these oppositions; it entails a realm in which their duality is no longer effective. Binarism in itself no longer obtains, and lost along with it is all opposition to an Other, an enemy; "Today it is increasingly difficult for the ideologues of the United States to name a single, unified enemy; rather there seem to be minor and elusive enemies everywhere" (189).

Simulation, spectacle, are everywhere in this global rhizome of connection and articulation. Commerce, communication, crisis, everything operates on a micro level, nothing is clearly localizable in a nation, a people, a soil, a city. The very dialectical operation (Marxism) by which such a movement could be analyzed no longer obtains: any dialectics is itself subsumed and rendered inoperative by a movement that generates and degenerates all play of oppositions. There are crises, indeed there are tiny movements of duality, of opposition, but these "minor and indefinite" crises add up only to an apparently inchoate "omni-crisis."

With the eclipse of dialectical opposition comes the death of nature in a very specific sense: with no outside, there can be no "natural world" acting on us, beyond our control. Hardt and Negri write:

> Certainly we continue to have forests and crickets and thunderstorms in our world . . . , but we have no nature in the sense that these forces and phenomena are no longer understood as outside, that is, they are not seen as original and independent of the artifice of the civil order. In a postmodern world all phenomena and forces are artificial, or, as some might say, part of history. (187)

This "postmodern" recognition of the end of nature throws us up against the one limit of this seemingly limitless, global skein of communication, commerce, exploitation, and wealth. If all is historical, all is, essentially, a function of labor. In this respect, Hardt and Negri remain fully in the tradition of dialectical Marxism, not to mention Kojève, for whom the death of Man really was the triumph of human labor in all domains: industrial, philosophical, historical. As Hardt and Negri put it,

> From manufacturing to large-scale industry, from finance capital to transnational restructuring and the globalization of the market, it is always the initiatives of organized labor power that determine the figure of capitalist development. Through this history the place of exploitation is a dialectically determined site. Labor power is the most internal element, the very source of capital. (208)

What is remarkable in all this is the occlusion of energy as a source of value. If energy, at least in a basic sense, is the power to do work, it too must have a profound connection with capital. It must in some sense be linked to productivity, inseparable from it. Yet Hardt and Negri never consider energy. Perhaps this is due to the fact that the source of energy, unlike the source of human labor, is not human. There is a finite quantity of oil available. Human labor did not put it there; Man was not involved. Humans extract it, to be sure—labor linked to machines that themselves consume fossil fuel energy (pumps, transport devices, etc.)—but the energy itself is derived from the consumption of depletable fuels. As Beaudreau would put it, human labor plays largely a supervisory role. Energy could not be tapped without labor, it is necessary, but it is hardly sufficient to the generation of capital. In this sense we could question whether labor is "the most internal element"—perhaps "one internal element" would be preferable (the other element being energy derived from inanimate sources).

I raise this issue of energy in order to resituate the argument of *Empire*. If for a moment we assume that the global world of commerce, replete with electronic media, the Internet, virtual television, and whatnot, is the replacement for and the simulacrum of the nonuniversal[21] city, we can only conclude that it can be so only as long as "nature no longer exists." But the fact that nature no longer exists, or at least seems no longer to exist, depends, ironically, on a natural given: the presence of fossil fuels in the earth—oil and coal, primarily. Labor power discovered these fuels, put them to work, "harnessed" them, transformed their energy into something useful. But labor power did not put the fuels in the earth. And perhaps more important from our perspective, it will be hard-pressed to replace them when they are gone. Nature-produced energy—the "homogeneous" energy that lends itself to work and the other, "heterogeneous" energy that is sovereign, not servile.[22] If the very term "nature" is contestable, one thing that cannot be contested is that the primary sources of energy come from natural sources: millions of years of algae accumulating in certain ecosystems, for example.[23] Thus pollution, dependent on this energy from natural sources, is ultimately natural; so too is global warming. So too is the incomprehensible unharnessed energy of the universe, which our labor and knowledge can only betray. So too will be massive die-off of humans and other organisms at the point of depletion. Man as the author of his own creation—homo faber—is opened by the radical exteriority, the finitude, the heterogeneity, but also the infinite richness of "nature." Man, as Sade would remind us, can never hope to have his reason domesticate a nature that "threatens the adequacy of rational systematicity"[24] or that defies the seeming necessity of all human activity. Nature deals death, and there is no way, finally, to grasp it by simply exploiting it ("knowing" it) as a resource or analyzing away its threat as sublime difference.

These considerations lead us to a much larger question: if Empire is dependent on fossil fuel energy for its simulacral effectiveness, what happens to it when energy becomes more expensive? When the "essence" of the human is no longer gasoline *(l'essence)*?

Without energy, there is no instantaneous communication, no movement of products and commodities across oceans, no $15 Wal-Mart sneakers shipped in from China. What is the energy input needed for the fabrication of a single computer workstation? For a single e-mail message?[25]

As energy inputs get tight, so Empire becomes Imperial. It becomes, in other words, once again central: lack of instantaneous movement, the depletion of the sheer now of virtual space, results in the reaffirmation of place, of the local. From fossil fuel Empire to the imperial realm of solar power, the energy of the local.[26] Things get a lot harder to move when only human labor power—along with animal power, perhaps—does the moving. Things get harder to grow when humans have to pull the plow and have to fertilize with their own excrement.

The global was the city of the empty *now,* of space as a two-dimensional blur. The automobile as universal and empty signifier. The statues fell, all was empty, instantaneous. There was a network of points, a lateral spread, but no limit to the "now" or to the quantity of information that could move. To be able to write these sentences in the past tense indicates the imminent return of time, history, change.

Depletion was there from the outset: in the first lump of coal burned, there was energy derived from the destruction of a resource. The fossil fuels, and the electricity, that drive Empire *are* depletion: if they could not be depleted, burned up, they could not provide energy. Fossil fuel is the passage of time as decay, as petrifaction, the shift from life to death as living things are transformed into valuable and exploitable reserves; it is also the ticking time of the loss of resources, of the limit. Empire in its seeming posthistorical timelessness, in its adialectical spread, tends nevertheless toward depletion, toward death as a wasting of the ideal figures of consumption, meaning, Man as speed. Empire necessarily, from the first, in its energy profile, implies its own extinction. For the statues to fall requires massive energy inputs. Waste as "practical"—all movement of Empire—depletes itself, reveals itself as the death, the emptiness at the heart of a fossil fuel globalization and the technoscientific realization. Man caught in his timeless location in Empire is ejected at the moment the economy fails: the moment of the exhaustion of resources, the point at which the growth economy falters because the ever increasing accumulation of debt, based on the seemingly endless increase in the availability of human-friendly energy (the power to "do work"), grinds to a halt.[27] The logic of growth is inseparable from the logic of depletion: finitude at the heart of the seemingly infinite.

And this fall teaches us of another incarnation of God, another incarnation of Man, because like God in the Middle Ages, or like Man in the twentieth century, the continued availability of fossil fuel energy today is an article of faith. It is almost impossible to speak of Hubbert's peak nowadays because people assume, for no good reason at all, that "it will never happen."[28] In medieval times not everyone was a believer, but everyone in some way took God's existence for granted. And the existence and centrality of Man was equally evident in the twentieth century. So too today, many of the experts and directors of our culture do not "believe in" fossil fuel energy—they know that it is finite, that "other sources must be found." But they cannot not believe in it either. So they go on, assuming its presence, that useful energy will always be there, will always be invisible, making effective work possible, effacing itself before the all-powerful Man, human labor, human creativity, the spirit of human invention, the human mind.[29] Man is the ultimate avatar of cheap energy. The death of Man—following its double, the death of God—is thus inseparable from the event of the finitude of fossil fuel. And the transgression of that finitude is nothing more than the affirmation of an intimate world, a world of the expenditure of, or "communication" with, another energy, one whose exile was necessary for the establishment of the dominion of energy that "does work."

When we recognize that human labor is not the sole source of value, and that energy slaves are indispensable to all the cultural, not to mention industrial, production we associate with modernity, then we confront the fact that everything must be rethought. "Growth" is not a function of a certain economy, of the borrowing and printing of ever more dollars, but of (for the time being) constantly increasing supplies of fossil fuel resources. How do we think the corollary of growth, retraction, in an era of depletion? Are democracy and liberalism separable from a rich fossil fuel mix? Women's rights, gay rights? Can philosophy in its various stages of development be directly linked to the fuel mix obtaining at different moments of history? What are the implications for philosophy of, say, a constantly rising or shrinking number, per person, of "energy slaves"? (Shrinking, perhaps, to the actual return of slavery.) What are the ethical implications of a massive die-off of billions of people?[30] It is tempting to assume a simple reverse dialectic: the return of the feudal. Thus one can anticipate the return of feudal (solar and slave) energy, feudal oppression (think of the Incas), feudal religion (literalism, the Book). But there is more than an energy blip that separates us from the Middle Ages: just as Bataille doubled Kojève, reenacting absolute knowledge, definitive reason, and then miming it, we may take for granted the knowledge that has evolved in modernity—its utility, its truth value, its radical limitations.[31] In that sense we need not assume that a downside of the culture curve will exactly parallel the downside of the fuel production curve. Such determinism is inevitably defeated by non-knowledge.

The fossil fuel regime nevertheless "falls" into the material, the particular, its knowledge into non-knowledge: the cursed matter of the body, the sacrifice of a comfortable particle enshrined in a fast personal vehicle. Negativity passes from an era of its putative full employment to an era of fundamental unemployment—"negativity out of a job," as Bataille dubbed it.[32] The homogeneous energy of the standing reserve falls into the heterogeneous, recalcitrant energy of ritual expenditure. Depletion and expenditure, Hubbert's peak and Bataille's peak, converge.[33]

Heterogeneous energy is *insubordinate,* not only as that which is "left over" and "unemployed" after the job is done, but above all as that which is a priori unemployable, always situated just beyond the limits of sense and growth. In this sense it is inseparable from the sort of "base matter" that cannot lend itself to constructive purposes or scientific edifice building, which Bataille discusses in such early essays as "Formless," "The Big Toe," and "Base Materialism and Gnosticism." Energy in Bataille is therefore dual, as is matter. Perhaps it is in this sense, and this alone, that Bataille can state (in *The Accursed Share*) that there are infinite quantities of energy in the universe. All that energy, however, is precisely *not* employable: it is energy that is burned off, that accomplishes nothing on a mere practical human scale. It transgresses the limits of depletion. In this sense one has to be wary of official definitions of energy that would hold that it is the "power to do work." The very use of the term "work," albeit in this context a purely neutral, scientific term, nevertheless does tend to anthropomorphize energy: its power not only, say, moves or heats: it accomplishes something in a purposeful, human sense. Similarly, does the expenditure of energy, on a cosmic scale, necessarily entail the transformation of a more "ordered" energy into a lesser "ordered" one? Ordered for whom?[34] Without the presence of a (human) subjectivity demanding the highly "ordered" energy, can one speak or order? Can it even be said to exist? In speaking of the finitude of energy supplies, we are only speaking of the limits to the human, the fundamentally limited availability of ordered energy capable of doing "work" for Man. We are speaking, in other words, of death, of the incommensurability of intimate Nature.

There is a limit to order: establishing order, fueling it, so to speak, but necessarily excluded by it, the infinite energy of the universe is a kind of black hole in which the ordered energy of creation is incessantly lost. The universe's energy is that of celestial and orgiastic bodies. "Heterogeneous" energy is nevertheless also the opening of the possibility of the distinction between useful and useless energy; it thus opens the possibility of work, but it cannot be subordinated to this distinction, which itself is useful (in opposing energy and entropy, utility and waste, sustainability and nonsustainability). It violates the limit of this distinction. (Once again the right hand is ignorant of what the left hand does.) The "tendency to spend" on the part of people in society is nothing more, as Bataille often reminds us, than the tendency to identify with the extravagance of the universe—the extravagance, in other words, of energy not subordinate to the dictates and "needs" of Man (the energy of the death of God and Man at the points of their greatest coherence). Energy is expended only in relation to Man: when Man is confounded with the universe, energy is neither expended nor conserved; it spends itself, we might say, opening but indifferent to the possibility of its use and waste. Indifference indeed: the indifference of ruins.

Depletion is not expenditure, but the limit beyond and against which expenditure plays. As we know, there can be no expenditure without a limit inscribed in a system: according to Bataille, a shortage of arable land, for example, is inseparable from the expenditure of goods and humans in the form of unproductive colonies of monks (the case of Tibet). It is the limit at the heart of the useful, the necessary, that opens the possibility of an expenditure without return. We could say that depletion is waste that is inseparable from the demands of necessity, that is conceived in and through necessity; but at the same time, its inevitability opens the way to the loss of sense, of bodily energy, of self, of God and Man. Its movement was perhaps necessary in the passage of knowledge into non-knowledge—a passage that in any case may be less a historical moment than an incessant movement, a fall. Or, put another way, that can be read as the circular agitation of sustainability and postsustainability, with sustainability the recurring aftereffect of the "night" of postsustainability.

The burning of fossil fuel in a sense makes this doubleness of depletion evident. We burn these fuels because we have to. We waste them because we have no other choice, it is necessary to our freedom, our destiny, or whatever. This senseless waste in its deepest intention is the establishment, outside itself, of what must be: wealth, Man, the security of our lives, our Law, and Meaning. Our nonnegotiable lifestyle. But this waste is not simply opposed to another expenditure because its heedless burning already introduces an irremediable loss at the heart of the necessary. A loss, a fall: from comfortable, driven Man to the straining muscles of the mover, the walker, the cyclist, the dancer. From the comfortable knowledge of Man to the risk of the human in a general economy. From the unconditioned as guarantee of meaning (God) to the unconditioned *as* unconditioned: monstrous, incomprehensible, joy and anguish before death. "Nature" in the violence of her transmutations, as the line transgressed to another energy, that of celestial bodies. Energy in and as human muscle power is not so much quantifiable or comprehensible, like a Foucauldian "bio-power," but is instead material (heterological, Bataille would say), the arousal, shuddering, and opening of organs and orifices—vehicles of transport—their communication with and in the night.

The radical finitude of fossil fuel—the Nature that refuses to die, even when it gives itself up and runs out (and its running out is its reaffirmation of its singular autonomy)—is the opening of muscle expenditure, the squandering of excited organs. The (im)possibility of the city as locus of *transport,* in all the senses of the word, in an era of postsustainability, entails a finitude inseparable from the movement of generosity, as among the gleaners in Varda's film.

What can we know of the future? Some would argue that, inevitably, the return to a solar economy is a return to a feudal economy, with all the exploitation and horror that that implies (Perelman 1981). The sun's energy harvested on Earth in its intensity can never match that concentrated in fossil fuel; since the energy provided by the sun is so much less, we can ultimately foresee the replacement of energy slaves by real slaves. Traditional slavery was merely a crude but effective way of appropriating an energy output from biomass (crops, wood), in order to free small numbers of people from the necessity of the expenditure of their own muscular energy: without engines, without appliances and the electricity that runs them, other people will again provide the energy. And most people, no doubt, will provide their own energy: this is called grinding poverty.[35] Many predict, as fossil fuel production declines, the breakdown of society, population die-off, the return of earlier social modes (Price 1995).

And a number of authors recently have been concerned with the methods we might use to avoid, forestall, or prevent this decline. (Odum and Odum 2001; Heinberg 2004). But others see a decline, a downfall, as inevitable

(Kunstler 2005b). No matter how we prepare, how we attempt to anticipate a time when there is little energy at the easy disposal of each citizen, we may very well face an intolerable decline in our civilization.

I would argue that while cheap energy is not the sole cause of the fabulous increase in population and wealth that the world has seen in the last century and a half, it is nevertheless inseparable from it. Bataille, in *The Accursed Share,* has indicated, correctly I think, the centrality of energy and its "uses" (or wastage) to the establishment and maintenance of any cultural formation. Likewise, the depletion of cheap energy will be inseparable from a return to a "sensation of time," to bodily expenditure (not least as "work"), and a charged, insubordinate matter. But at the same time it is the very "cursedness," the unknowability, of this matter that prevents its incorporation in a simple reverse dialectic. We cannot simply flip over the dialectic and predict a decline where previously there had been an advance. Our right hands don't know. The death of Man, of the certainties of Law and hierarchy, the atheism of God himself, the *known* finitude of "useful" energy, indicate that the unknowable future will not be conceivable as the simple downside of a bell curve, the simple disarticulation of a social dialectic, moving us backward from globalization to monopoly capital, from there to robber baron exploitation, from there to feudalism, centrally organized agriculture (employing slaves), and then maybe even back further, to a hunter-gatherer society. We cannot assume that we will be forced into any given social regime by any given energy regime. The moment of depletion is the fall of knowledge, of utility, of meaning "built up" through successive certainties. It will not necessarily entail a fundamentalist unity of God and creation (as I have tried to show), or a rationalist governance in which every "tendency to expend" is analyzed, known, and controlled for the benefit of Man. What separates the downside of the bell curve from the upside, then, is not only the refusal of any easy prediction, but also the fact of knowledge. Non-knowledge is not simple ignorance, but the following-through of the consequences of attained knowledge; in our case, this implies a full understanding of the energy regime of modernity, the benefits and pitfalls of rationalism and humanism. Reason, as applied to the understanding and governance of society, will not simply be forgotten. As we saw in *The Accursed Share,* the highest knowledge of society is the consequence of following through a reasoned analysis and understanding of societal drives, both rational and profoundly irrational, to their end. By the same token, however, non-knowledge means the impossibility of predicting a practical future whose sole beneficiary or victim is Man. A society that recognizes that the ultimate signified (God, Man) is heterogeneous in the most basic sense of the term is one that recognizes that its own impulses are both inescapable and profoundly gratuitous.

Following Bataille, we can argue that the future, the fall into the void of certainties (God, Man, quantifiable and usable energy) may lead to another kind of spending, "on the scale of the universe," which, in spite of itself, would entail what I have called postsustainability. We do not know; what is clear is that one kind of matter, one energy, one plenitude, is dying; another, monstrous, already here, already burning, announces itself. Hubbert's peak announces it, yet betrays it, for Hubbert envisaged only one version of energy. Up until now the development of thought, of philosophy, has been inseparable from the fossil fuel–powered growth curve, from "civilization." The downside of the bell curve is non-knowledge because the event of the decline of knowledge, the disengagement of philosophy from economic and social growth, cannot be thought from within the space of knowledge growth (the perfection of modern truth) or its concomitant absence. We are in unknowable, unthinkable territory—an era of disproportion, as Pascal might call it. The era of Bataille's peak.

"I love the ignorance concerning the future," wrote Nietzsche, and Bataille seconded him. For Bataille, any assurances concerning the future, either good or bad, were beside the point, even silly; instead, there was the play of chance, the affirmation of what has happened, what will happen. The left hand spends, in gay blindness as well as science, and the future is affirmed, in the night of non-knowledge.[36]

Does this mean that we should despair, and use this "ignorance" as an excuse to do nothing? Not at all; we know the difference between sustainability and catastrophic destruction; we know the difference between global warming and a chance for some, even limited, species survival. But we also recognize, with Bataille, the inseparability of knowledge and non-knowledge, the tilt point at which, rather than cowering in fear, we throw ourselves gaily into the future, accepting whatever happens, embracing everything, laughing at and with death. We will a return of recalcitrant bodily and celestial energy, of the sacrifice of the logic of the standing reserve; we bet against the vain effort to will an endless autonomist freedom. We know that sustainability, if such a thing ever were to come about, would be inseparable not from simple calculation and planning but from the blowback of the movement of an embrace of the transgressed limit, the intimacy of the world willed to ritual *consumation,* the embrace of death-bound bodies: post-sustainability.[37] In other words, after Bataille, we refuse to take the downside of the bell curve as a simple and inevitable decline into feudalism, fundamentalism, extinction. We understand all that depletion implies, and we embrace it, affirming the movement of expenditure at its Varda-esque heart.[38]

Who is this "we"? Not the self-satisfied "we" of a closed community or multitude, jealous of its rights and serene in its self-reflection. Rather, a not-we, emptied of meaning, unjustified—a community of those with nothing in common (Lingis 1994).

Notes

Introduction

1. See Fukuyama 1992 for a recent take on the Kojèvian model of the end of history. Kojève's version of Hegel will be discussed in chapter 3.

2. Gerald F. Seib, "Oil Dependency Overshadows US Policy," *Wall Street Journal,* 22 August 2005, A2.

3. On the inevitable connection between the depletion of fossil fuel resources and die-off, see especially Price 1995 and Zabel 2000. It should be recalled that the green revolution would have been impossible without the production of fertilizers derived from petroleum and natural gas.

4. See Heinberg 2004, 40–41.

5. True, there are other possible (renewable) sources of energy, which no doubt will be developed over the course of the present century. Many authoritative commentators (such as Heinberg 2003), however, note that the amount of energy to be derived from these alternative sources will never be as great as that provided by the fossil fuels, especially oil and natural gas (coal may continue to be plentiful, but deriving energy from it will prove more and more problematic, given issues of pollution, global warming, and even depletion).

6. Hence the title of Huber and Mills's book, *The Bottomless Well* (2005).

7. These alternatives—environmentally aware literalist (or evangelical) Christianity versus ecoreligion—are discussed in chapter 6.

8. On the two opposed varieties of the sacred, see above all Bataille's essay "The Psychological Structure of Fascism" (in *œuvres Complètes* [1970–88; hereafter cited as *OC],* 1: 339–71; and in *Visions of Excess* [1985; hereafter cited as *VE*], 137–60).

9. I am thinking here especially of Perelman's article (1981), which posits as necessary and inevitable the return to some version of feudalism when "modern" industrial society reverts to renewable, in particular solar, energy.

10. Many eco-economists hold that happiness depends not on the simple consumption and disposal of goods but on other intangibles such as membership in a community, strong social ties, and the feeling that one's work is worthwhile. For this reason, a future society, reorienting its priorities accordingly, can promise greater happiness for a larger number of people with much less consumption and

waste. (See, for example, Costanza 2006; and Mulder, Costanza, and Erickson 2005). While I do not dispute the fact that one can't buy happiness (the gist of the eco-economists' argument), I would argue that, following Bataille, there is indeed a "tendency to expend" that is at the basis of any vibrant community. What needs to be rethought is not only happiness but also expenditure. That's where Bataille comes in.

1. Bruno, Sade, Bataille

1. See, for example, Yates 1964, 264–65; while ascending all the way to the "Optimus Maximus," "incorporeal, absolute, sufficient in itself"—unconditioned, in other words—the magician must pass through the demons, and, it seems, they constitute a stopping point virtually autonomous in relation to God (the Optimus Maximus). As Yates writes, "Obviously a vital stage in the ascent would be to reach the demons, and Bruno's magic is quite obviously demonic.... Bruno wants to reach the demons; it is essential for his magic to do so; nor are there any Christian angels within call in his scheme to keep them in check" (265).

2. There was already a tradition of empiricism in European philosophy, dating to the period just before the Black Death, in the fourteenth century. Thinkers such as Duns Scotus and William of Ockham attempted to separate revealed wisdom and empirical observation, on the assumption that rational investigation or interrogation was of little use in the case of revelation. See Gottfried 1985, 31, 153. Bruno, on the other hand, was never able to separate his materialistic science from mysticism. As Stanley L. Jaki, in his introduction to Bruno's *The Ash Wednesday Supper* (1975) writes: "In particular, [Bruno] was tragically mistaken about science. For him science was the wave of the future only inasmuch as it served the cause of Hermetism, a synthesis of occultism, magic, cabbala, necromancy, and weird mysticism" (24). With this said, one can argue (as we will see) that neither Sade nor Bataille was able to separate entirely materialism from a "weird mysticism" either. And at this late date their critique of science does not seem that far-fetched.

3. On Bruno's clash with the church, see M. White 2002.

4. Again, however, the religious and the mystical in Bruno are never *exclusively* associated with a "right-hand" God of creation and redemption; in this sense, even with a "religion," Bruno could be said to anticipate a Bataillean "religion" of the death of God. Yates in *Giordano Bruno and the Mystical Tradition* (1964) writes: "The ultimate object of the magic memory was the formation of the religious personality, or the personality of the good Magus. Hence, after the mysteries of the thirty mnemonic 'seals,' Bruno enters upon a discussion of religion. This he does under the heading of different 'contractions,' by which he means different kinds of religious experience, some of which are good and some bad" (271). Note that, as with Bataille, but without explicitly posing the question of the death of God, Bruno poses an ambivalent religious experience, both divine and demonic—of the right and left hands.

5. Hilary Gatti (1999) says of Bruno's atomism: "If each minimum atom contains all the maximum power of God, then all matter is imbued with the absolute power of the divinity and finite only according to the subjective limitations in space and time of the perceiving qualities of the human mind" (114). This is a nice solution to the problem and would seem to get God off the hook; if we see matter and its energy—its divine "power"—to be as much destructive as constructive, that is only because of our own limited, finite viewpoint. This would seem to institute a radical division once again between human (finite) and divine (infinite) creation, the very kind of distinction Bruno was at pains to reject: his embrace of Copernicanism, after all, was an embrace of a universe that was consistent, free of the Aristotelian distinction between a higher, divine reality and a lower, finite one. Now that distinction is established in every bit of matter: matter is both infinite in its extension, in its unity, but finite in its individual instances. One notes here as well that for Bruno "infinite" means that God "exclud[es] from himself every limit" (113)—a quality, if one can call it that, that Bataille also recognizes when he considers that God is radically unconditioned, sovereign, "accountable to no one." But this radical freedom for Bataille implies not a God who is universal, consistent, wise and good in his infinity, but rather a dead God, inseparable from the chthonian forces of destruction. The distinction between infinity and sovereignty is the distinction between the left- and right-hand sacred, of course; but if one refuses any hierarchy, any easy distinction between God and creation, then one moves from the infinite as consistent, good, etc., to the sovereign as base and violent: to the radical finitude of death, in other words, in and with which God is complicit, inseparable.

6. As Denis Hollier puts it, "Evil does not exist independently of the interdiction which is the limit of the Good; beyond this limit reigns only another Good; not Evil. Evil never reigns. . . . In Gnostic dualism, Bataille was seduced by a contradictory materialism which he opposed to the physicists' mechanical and rational materialism which, because it is monistic, he called a 'doddering idealism.' Thus Bataille's attitude can be portrayed as a *dualist materialism:* an 'impossible' attitude, as distant from theology as it is from even an atheistic humanism which he named atheology" (1990, 130).

7. Morality, for d'Holbach, consisted of the rational principles, gathered through sense experience, necessary to the successful functioning of society and to our happiness within it (Vitzthum 1995, 74–77). Sade's countermorality, ostensibly devoted to freedom from tyranny (from harsh laws) is itself, of course, equally dependent on sense experience. The highest good, both for d'Holbach and Sade, is sensual pleasure.

8. Caroline Warman holds that Sade in effect rigorously rewrites earlier eighteenth-century materialists (d'Holbach, Helvétius): "By 'materialist derivation' I mean that the source for both plot and character [in Sade] is provided by the theories of the behavior of matter current at that time" (2002, 157).

9. On materialism and the mystical-alchemical tradition in Sade, see ibid., 108–14.

10. On *creuset* (crucible), see ibid., 109.

11. Nevertheless, eighteenth-century materialists held that a coherent morality was possible on the basis of materialism. Here is the one area in which they did differ radically from Sade. Whereas Sade's morality, if one can call it that, is one of what typically might be called crime or evil—orgiastic murder is celebrated as the highest good—materialistic morality for thinkers such as d'Holbach is a more perfect version of Christian morality. See Vitzthum, for example, on d'Holbach's morality: "Chapter 12 [of d'Holbach's *System of Nature*] has the impossible job of arguing that human beings are accountable, despite their lack of freedom, to a single moral standard valid in all circumstances" (1995, 75). A moral standard, needless to say, that would promote charity and be "useful to the whole human species"—something Sade, of course, was notably indifferent to. Interestingly enough, when d'Holbach comes closer to Sade's concerns—an attack on religion—Vitzthum finds it "stingingly successful" (81). One has the sense that, given the difficulty of establishing a rigorous affirmative morality and the ease with which one can dismantle religion, eighteenth-century materialism ultimately lends itself much more easily to a Sadean philosophy than to a d'Holbachian one.

12. See my analysis of Agnès Varda's film *The Gleaners and I,* in chapter 5.

13. See *La Philosophie dans le boudoir* (1968): "We should never impose on the murderer any other penalty than the one he might risk from the vengeance of the friends or the family of the one he has killed" (270; my translation).

14. The "left-hand" sacred is another term for the "impure" or "unlucky" sacred. This distinction originally was proposed by Robertson Smith; Durkheim develops and analyzes the duality of the sacred in *The Elementary Forms of the Religious Life* (1995, 412–17). For Bataille the sacred that counts is always the left-hand sacred, the one that logically predates (and that is blind to) the recuperative action of the right-hand sacred. The latter is inseparable from a religion that betrays religion: that proffers an unconditioned (sovereign) godhead who nevertheless serves as a guarantee to human existence and as meaning and (deferred) end for all human activity. The left-hand sacred has the immediacy of the now; it is its own end; it is an electric shock—the jolt of unrecuperable energy—in the present. Left-hand sacred objects are excluded, rejected as base, horrifying, nauseating. The right-hand sacred is that of temporality, of the harnessing of energy to a future goal, to an (indefinitely) projected end, to a larger, legitimizing meaning. Right-hand objects are elevated, pure, holy, untouchable in their goodness. Both the left- and right-hand sacreds are, strictly speaking, "heterogeneous," useless, they do no work, and do not enter into the ("homogeneous") economy of balanced accounts; the difference, however, is that the right-hand sacred lends itself to, gives meaning to, serves as the end of, useful and goal-oriented activity; the left-hand refuses this service. To put it crudely, the right-hand sacred, typified by an absolute and elevated God, has sold out.

15. As Durkheim puts it, "Religious forces are in fact only transfigured collective forces, that is, moral forces; they are made of ideas and feelings that the spectacle of society awakens in us, not of sensations that come to us from the physical

world" (1995, 327). At the same time these forces are analogous to an electrical charge in that they pass from charged object to object, on contact: "Thus, the heat or electricity that any object has received from outside can be transmitted to the surrounding milieu, and the mind readily accepts the possibility of that transmission" (327). In effect, for Durkheim this energy is a social one entailing both the institution and the self-representation of a social group. On this topic, see chapter 1, "Durkheim and the Totem-Act," of my book *Agonies of the Intellectual* (1992). It is not scientifically analyzable (outside of the science of anthropology): it behaves like electrical energy, but it is not electricity.

Bataille for his part rejects any desire to represent the force of the "heterogeneous" as somehow objective, outside of a collective movement in which the certainty of the human (science, philosophy) is overthrown. In "The Use Value of D. A. F. de Sade," he writes: "The heterogeneous is even resolutely placed outside the reach of scientific knowledge, which by definition is only applicable to homogeneous elements. Above all, heterology is opposed to any homogeneous representation of the world, in other words, to any philosophical system" (*OC,* 2: 62; *VE,* 97). Bataille's difference from Durkheim here is that while Durkheim sees the analogy of *mana* and energy, with *mana* serving the useful function of unifying groups and society as a whole, Bataille sees its "equivalent," the heterogeneous, as actually threatening any coherent structures that would establish or maintain the philosophical or scientific truth upon which a developed modern society rests.

16. "Fiction" implies, in contradistinction to it, Truth; myth, on the other hand, like tragedy, is inseparable from the virulent movement of energy (in crowds and body movement, in the energy of "celestial bodies," in the movement of base matter) that makes possible the opposition of truth and fiction (or useful and useless energy) but is not reducible to it. Energy, in other words, can lend itself to servility in sense, in survival, but as a larger movement it cannot be so tamed.

17. Thus in *Erotism* (first published in 1957) Bataille writes of the "thieving rabble" and prostitutes: "Prostitutes fall as low as they do because they acquiesce in their own sordid condition. That may happen involuntarily, but the use of coarse language *looks like a conscious decision;* it is a way of spurning human dignity. Human life is the Good, and so the acceptance of degradation is a way of spitting upon the good, a way of spitting upon human dignity" (*OC,* 10: 137; *Erotism* [hereafter cited as *E*], 138; italics added). Although Bataille is not as positive here concerning this class as he was in the prewar writings, we see the same model: degradation, the violation and "dragging in the mud" of the Human, *reaffirmed consciously* (therefore distinct from sheer animality).

18. Hence Sade's utopia: "Various establishments, clean, vast, correctly furnished, secure in every way, will be erected in all the towns; there, all sexes, all ages, all creatures will be offered to the whims of the libertines who will come to enjoy *[jouir],* and the most total subordination will be the rule of the individuals presented; the slightest refusal will quickly be arbitrarily punished by he who experiences it. I must continue to explain this, relating it to republican morality" (Sade 1968, 232; my translation).

19. Hence Bataille's interest in creating mythical figures such as the "acéphale," the headless being who represented not simply a fanciful or false creation, but a heterogeneous force in the world, imbued with the violence of the left-hand sacred. Mythical figures would carry, in other words, the virulence, the fundamentally undisciplined energy, of "mana"; in this sense they would hardly be imaginary or fictional. They would partake of the insubordinate aspect of base matter or excess, unusable energy (be it the energy of the body or of the stars).

20. For Bataille the workers' experiences of expenditure—that which truly "liberated" them—was inseparable from the violence of revolution itself, revolution as potlatch, in other words. As he writes in "The Notion of Expenditure," "In historical agitation, only the word Revolution dominates the customary confusion and carries with it the promise that answers the unlimited demands of the masses.... Class struggle has only one possible end: the ruin of those who have worked to ruin 'human nature'" (*OC,* 1: 318; *VE,* 127–28).

21. In "The Pineal Eye," Bataille writes: "From the first, myth is identified not only with life but with the loss of life—with degradation and death. Starting from the being who bore it, it is not at all an external product, but the form that this being takes in his lubricious avatars, in the ecstatic gift he makes of himself as obscene and nude victim—and a victim not before an obscure and immaterial force, but before great howls of prostitutes' laughter" (*OC,* 2: 25; *VE,* 82).

22. In Lacanian terminology, this science would be concerned with the real. See van Wyck 2005, 96–103, 112–14.

23. Mauss in this theory of the gift in fact anticipates Bataille, specifically in his rejection of selfishness in the establishment of a gift-based economy: "Social insurance, solicitude in mutuality of co-operation, in the professional group and all those moral persons called Friendly Societies, are better than the mere personal security guaranteed by the nobleman to his tenant, better than the mean life afforded by the daily wage handed out by managements, and better even than the uncertainty of capitalist savings" (1967, 67). One notes, at the same time, a profound difference between Mauss and Bataille: for Mauss these "friendly societies" guarantee social stability and security; for Bataille, on the other hand, gift-giving or destruction entails the anguish of the opening of the self to erotic torment or death. The critique of personal security for Bataille entails much more than the "mutuality of co-operation."

24. An autobiographical project, we might add, but autobiography as the sacrifice of the coherent authorial self (*Inner Experience;* hereafter cited as *IE*).

2. Bataille's Ethics

1. Hochroth's superb article is necessary reading for any understanding of Bataille and energy. She very cogently traces out Bataille's approach as a critique of Carnot and as a reading—and political critique—of Wilhelm Ostwald. Both Bataille and Ostwald are energeticists, Hochroth notes. Energeticism proposes that "all occurrences are produced by differential intensities, otherwise known as the law of

transformation. In this way, as opposed to the mechanistic view, everything can be explained in terms of energy" (1995, 71). The difference between Bataille and Ostwald is between their respective "politics of expenditure"—a difference leading to a "different scale of values" (73). Ostwald believed not in glorious expenditure but in "reducing waste to a minimum" (76).

2. See on this topic Diamond 2005 and Manning 2004.

3. There is a tendency among recent critics to consider the problem of energy, its conservation and expenditure, as somewhat outdated; see Clarke 2001, for whom the question of energy is replaced by that of information.

4. See Huber and Mills 2005.

5. Recall that for Bataille the sacred is double, both the elevated, conservative sacred of established and productive religion (the right-hand sacred), and the base (left-hand) sacred of cursed matter and orgiastic, lawless loss.

6. In any case the external and internal limits of a society are closely linked. The catastrophic result of a social system that cannot conceive of—i.e., theorize—its limits is in fact the central question of *The Accursed Share.* Such a society still has to face its excess but, doing so in ignorance, it risks destroying itself. Such is the danger of the cold war face-off between the United States and the USSR—or so Bataille would argue—and such is the danger facing a society that cannot theorize its ecological limits (so I would argue, following Bataille's lead).

7. One can of course make the same generalization about animal ecologies; the struggle between animals, and even the internal limitation of animal populations in some species, takes place when energy supplies are scarce. In the case of human societies, warfare over limited resources and land is an example of this kind of conflict.

8. Derrida characterizes the crucial fifth section of *The Accursed Share,* in which nuclear destruction and the Marshall Plan are discussed, as "most often muddled by conjectural approximations" (Derrida 1978, 337n33). One can argue, in fact, that the vast majority of readings of Bataille tend to downplay or dismiss the social, ethical, and political implications of "general economy" and see it instead as a critique of Hegelianism or metaphysics writ large (ibid.), as a critique of epistemology (Hollier 1990), or of modernist aesthetics (Bois 1997). My argument, on the contrary, is that to a large extent Bataille was a social and even utopian (or dystopian) thinker whose vision of the future entailed a radical alteration in (the study and practice of) economics, religion, and eroticism. Bataille even called it "Copernican" at one point, no doubt alluding not only to the importance of the sun in Copernicus's theory but to the Copernican tradition, starting, of course, with Giordano Bruno.

9. See, for example, Foucault's *Discipline and Punish* (1995)—especially the importance he places on the (constitutive) role of violent spectacle in society—and Lingis's emphasis, for example, in *Trust* (2004), on transgression and excess in interpersonal relations.

10. *OC,* 7: 40; *The Accursed Share* (1988a; hereafter cited as *AS*), 33–34.

11. In this sense "Man" for Bataille is exactly *opposed* to a humanist Man who conserves and works, jealously guarding his property, life, and the sanctity of his

God (whom he creates in his own fearful image). This is the Man not of constructive labor and the desire for tranquility, but Man the painter of the Lascaux caves (see chapter 6). A "primitive" Man, then, who practices a religion and an economy "on the scale of the universe."

12. "Carrying capacity" refers to the population (of any given species) that a region can be reasonably expected to sustain. It is defined by LeBlanc in this way: "The idea [of carrying capacity] in its simplest form is that the territory or region available to any group contains only a finite amount of usable food for that group. Different environments can carry or support different numbers of people: deserts can support fewer people than woodlands, the Arctic can support very few, and so on" (2003, 39).

13. I should stress in this discussion that "expenditure" in Bataille's sense *(la dépense)* is a term to be valorized only in relation to the human. After all, the stars, in their burning of inconceivable amounts of hydrogen, do not expend it because there is no purpose to which it can be put. "Expenditure" is a term, like "transgression," that has meaning only against another term. One can only "transgress" an "interdiction"; one can only "expend" in relation to a need, or command, to conserve. Man's identification with the universe as expenditure is only in the sense that spending like and with the universe puts into question the integrity and coherence of Man. The energy of the universe is excessive precisely because it cannot be put to work by Man. Man's "tendency to expend" is therefore a profound movement that signals the very demise of the category "Man." Energy is only energy—the power to do work—in relation to Man. Energy is only loss in relation to the death of Man. A more profound energy—that which cannot be harnessed, that which is spent and not conserved—can only be said to be expended against the limit that is the finitude (and meanness) of Man. But if we were not human we could not write of these things at all. A total identification with the universe would result not in Bataille's writing, but in the vast undifferentiated, powerful, but purposeless energy of the black hole.

14. My translation; Ambrosino's text is not included in the English translation of *The Accursed Share.* Many of Ambrosino's texts of the late 1940s and early 1950s published in the review edited by Bataille, *Critique,* display the same assumption, that energy is available in infinite supply—not only in the universe as a whole, but in modern fossil fuel–based economies. This is a most peculiar position for a trained physicist to take. See, for example, his review of Norbert Wiener's *Cybernetics,* in *Critique* 41 (October 1950): "Energy, in a physical sense, is everywhere (the least gram of matter, etc. . . .), the sources of negative entropy, with which man furnishes himself, and his industry, are practically inexhaustible *[intarissable]*" (80).

15. On EROEI and its implications for any energy retrieval, distribution, or consumption system, see Heinberg 2003, 138. Another way of looking at this issue in Bataille (if not Ambrosino) is to think of two kinds of energy: an energy in fuel that is easily stockpiled, easily accessed, and another energy, elusive, difficult to capture and to use, an energy that is a priori excessive, active before the buildup of fossil fuels, in excess after their depletion. This is a larger energy, neither useful nor

wasted in the way that fossil fuels are wasted: energy as excess, in excess, energy as the defiance and mortality of Man (what is there before him, what makes him possible as energy consumer, but what eludes him as well).

16. The last really massive oil field found by petroleum geologists was the Al-Ghawar, discovered in Saudi Arabia in 1948 (one year before the publication of *The Accursed Share*).

17. See Heinberg 2003, 142–46.

18. Huber and Mills, conservative "infinite energy supply" critics, celebrate the "high grade," "well ordered" power that is derived from much larger quantities of "low grade energy." As they put it, "Energy consumes itself at every stage of its own production and conversion. Only about 2 percent of the energy that starts out in an oil pool two miles under the Gulf of Mexico ends up propelling two hundred pounds of mom and the kids two miles to the soccer field" (2005, 51). What Huber and Mills tend to ignore is that the fuel that allows this refining and concentration into the form of high-grade power is extremely limited. There are not a lot of docile energy sources that allow themselves to be quantified and put to work, and they affirm, in their combustion, our own finitude, our own mortality (they "serve" us only through their own depletion, through their acceleration of the arrival of the moment when we will be left without them). Most energy sources instead call attention to human finitude in that they are *insubordinate to the human command to become high-grade power.* They are the affirmation of the fundamental limitation in the quantity of available refinable energy sources; this is energy that does not lend itself to simple stockpiling and use. It is an energy with a sacrificial component, coursing through the body, parodically moving people through the city as they gawk and display (the walking of the flaneur), "transporting" both the body *and* the emotions. If this energy is infinite, it is so in the overwhelming quantity that is lost to us and not in its availability to our hubristic attempts at harnessing it.

19. See Beaudreau 1999 (7–35), for whom value in industrial economies is ultimately derived from the expenditure of inanimate energy, not labor-power. Conversely, "[human] labor in modern production processes is more appropriately viewed as a form of lower-level organization (i.e., supervisor)" (18).

20. It has taken millions of years of concentration in the fossilization process to produce the amazingly high-energy yields of fossil fuels: tapping into sunlight alone cannot come close. The sun is fundamentally resistant to human attempts at harnessing it; its power is of the moment, not of quantified stockpiles. F. S. Trainer, for example, sees enormous problems with the use of solar energy to fuel human society, even the most parsimonious: the difficulty of collecting the energy in climates that have little direct sunlight (1995, 118); the inefficiency of converting it to electricity and storing it, where at least 80 percent of the energy will be lost in the process (118–24); and even the expense of building a solar collection plant where, in Trainer's estimation, "it would take eight years' energy output from the plant just to repay the energy it would take to produce the steel needed to build it!" (124)—all these facts indicate that solar energy in relation to human civilization is, well, too diluted.

21. The "energy slave" is based on the estimate of mechanical work a person can do: an annual energy output of 37.2 million foot-pounds. "In the USA, daily use per capita of energy is around 1000 MJ, that is, each person has the equivalent of 100 energy slaves working 24 hours a day for him or her" (Boyden 1987, 196).

22. The dependency on concentrated sources of energy might be alleviated by ever more sophisticated and efficient methods of energy production and use (e.g., wind and solar energy, cellulose-based ethanol), but it can never be eliminated because none of the alternative sources of energy promise anywhere near as great an EROEI as the fossil fuels. See above all Heinberg 2003, 123–65.

23. See Odum and Odum 2001, 131–286.

24. See Diamond 2005 and Manning 2004.

25. See the prime example cited by LeBlanc: an area of Turkey where he did research as a young anthropologist. "Almost 10,000 years of farming and herding have denuded an original oak-pistachio woodland, and today [in a photograph of the area] only a few trees can be seen in the distance" (2003, 140).

26. Perhaps complimentary copies of *The Accursed Share* could be distributed in the Department of Defense and the Pentagon.

27. Heidegger's analysis of the defamiliarization, so to speak, of the useful but unremarked object in *Being and Time* clearly anticipates Bataille's take on the object, the "thing," both useful (and hence largely invisible) and in another, "general" context, not. For Bataille, the object is not just broken, wrenched out of its familiar context: it is rendered orgiastic, insubordinate, "cursed matter," etc., here following the tradition of reading the "sacred" in the French anthropological tradition (Durkheim, Mauss).

28. In Heidegger's famous postwar essay, *The Question concerning Technology* (1977). See my discussion of Heidegger in Chapter 5.

29. It can be argued that he did make the distinction, in "The Notion of Expenditure," by distinguishing between consumption *(consommation)* and what I would translate as burn-off *(consumation)*. Already in 1933 Bataille saw that the consumer society was not engaged in a kind of glorious, high-tech potlatch. It would appear that Bataille then forgot about this distinction later, in *The Accursed Share,* when he identified the Marshall Plan as a modern version of gift-giving (analogous to potlatch). Bruno Karsenti, discussing Bataille in the context of Mauss, makes the point clearly: "One must avoid, in this matter, confusing Bataille's perspective with that of sociologists who, like Halbwachs for example, first conceive of expenditure *(la dépense)* by relating it to the phenomenon of consumption *(consommation)*. For Bataille, consumption, taken in the strictest sense, is never more than a 'means to the end of production' ("The Notion of Expenditure" [*I,* 305; *VE,* 118]). It always has to do with quantification and measurement and does not obey fully the principle of loss. It remains linked to the superficial logic of utility. Its meaning is consequently weaker than that of burn-off *(consumation),* the pure expression of sociality" (Karsenti 1997, 444n2; my translation).

30. Technically energy is the "power to do work," but despite scientific dis-

claimers, "work" is obviously an anthropocentric notion. No "work" is done beyond the confines of the human community. From Bataille's perspective, the energy of the universe, the sun, and "celestial bodies" does no work, can do no work—and that is why it is infinite or (put another way) insubordinate to human needs and demands. Surplus energy on the earth, energy that cannot be put to work (for whatever reason), partakes of this "sovereign" aspect of the universe's energy. Thus one could speak of a "heterogeneous" energy distinct from the "homogeneous" energy of work just as one can speak of a heterogeneous (base) matter distinct from homogeneous (useful, elevated) matter (as Bataille does in "The Psychological Structure of Fascism," *OC,* 1: 339–71; *VE,* 143–48). It should be noted that this "heterogeneous" energy cannot simply be opposed to an energy put to use, stockpiled, and quantified for human purposes. Rather, the energy of the universe, if we can call it that, is neither useful (used) nor useless (wasted), but its excessiveness opens the possibility of this distinction (used/wasted). In a sense it is prior to the opposition between the two energies and escapes their opposition. To partake in this energy—if we can use this term—is to transgress the opposition inseparable from the institution of sense (to use stockpiled matter or energy; to waste it). Heterogeneous energy, like base matter, subverts oppositions without being incorporated back into them—except when, of course, we attempt to understand and thus inevitably betray these elements. Hence the evident paradox (and necessity) of a "heterological science."

31. This is the legacy of a well-known article by Jean-Joseph Goux (1990a), which I will discuss in chapter 5. See also Susan Blood's informative and useful article on Bataille and Derrida in this context (2000).

32. In "The Notion of Expenditure," Bataille writes: "The consequences [of potlatch] in the realm of acquisition are only the unwanted result—at least to the extent that the drives that govern the operation have remained primitive—of a process oriented in the opposite direction" (*OC,* 1: 310; *VE,* 122). Since the gift cannot be given with an end in view—in that case it is not a gift—there are only two other possibilities: (1) There is no such thing as a pure gift, giving is always a question of self-interest, and any theory of "spending without return" is sheer illusion; (2) gift-giving occurs in an intimate world in which there is a (Nietzschean) radical ignorance of the future—or, better, in a world where the calculus of planning and deferral does not obtain. This would certainly seem to be the case for Bataille; Derrida also suggests this when writing of the New Testament proposal that the left hand remain in ignorance of the right. As Derrida puts it in *The Gift of Death:* "Does this commandment [the ignorance of the right hand in relation to the activities of the left] reconstitute the parity of the pair rather than breaking it up, as we just suggested? No, it doesn't, it interrupts the parity and symmetry, for instead of paying back the slap on the cheek . . . one is to offer the other cheek" (1995, 102). See also Horner 1995, 219.

33. I use scare quotes advisedly here. This is not a Christian salvation or even the conservation entailed in, say, the "saving" of life. It is rather the maintenance of

a livable and mortal world, projected as far into the future as possible (a future made possible through a certain recycling, control of global warming, etc.), as an after-effect of an affirmed general expenditure.

34. See Onfray 2005 for a trenchant and rousing critique of the cultural bases of the monotheistic religions (the "religions of the Book").

3. Bataille's Religion

1. It should be recalled that for Bataille the term "religion" (or "religious") could be applied to social moments of the left-hand sacred as well as to those of the right. Hence Bataille considered the Acéphale group to be "religious," although it is of course not religious in the same way as, say, the Catholic Church. "We are ferociously religious," Bataille writes in 1936 ("The Sacred Conspiracy," *OC,* 1: 443; *VE,* 179).

2. As Blanchot puts it in *Faux pas:* "We speak of it as an experience and, nonetheless, we will never say that we have experienced it. Experience that is not a lived event, even less a state of ourselves, at the very most the *limit-experience* or, perhaps, where the limits fall" (1971, 51). On the experience as a nonexperience, see also Sichère 1982.

3. Bataille would not include cults like the gnostics in this characterization of religion since this "doctrine" explicitly entailed (according to Bataille, at least) an elevated, eternal God, creator and principle of the universe. The gnostics, instead, practiced a "left-hand," "impure" religion.

4. On the practice (if one can call it that) of Bataille's mysticism, as opposed to that of Christian (in spite of everything . . .) mystics like St. Angela of Foligno, see Hollywood 2002. On mysticism and the act of writing in Bataille, see Connor 2000. Both Hollywood and Connor provide excellent in-depth analyses of what is arguably *the* question in Bataille: the relation between religious experience and the project (labor, symbolization, etc.).

5. One thinks of Sartre's criticism of Bataille (in "Un Nouveau mystique"): "But if the joys to which M. Bataille invites us only lead back to themselves, or if they cannot be reinserted into the order of new enterprises, contributing to the formation of a new humanity that goes beyond itself on the way to new goals, they are worth no more than the pleasure of having a drink or relaxing in the sun on the beach" (1975, 228). (*Pace* Sartre, given the horrors of the twentieth century, having a drink on the beach—a perfectly wonderful form of unproductive expenditure—does not seem like such a bad option.)

6. The irony here, of course, is that Bataille borrows/parodies the title of St. Thomas of Aquinas's great work, the *Summa Theologica,* a work that ties the Christian tradition directly to the philosophy of Aristotle—a classical philosophy devoted to what can be known rationally.

7. Blanchot makes the point that, finally, the privileged term in Sade is not Nature, which itself is ultimately defied, but *energy.* The Sadean hero is above all a supreme concentration of energy, to the point of profound apathy, to the point of the death of the human. See Blanchot 1963, 42–46.

8. Exactly the same kind of progression can be seen, on the level of argumentation, in the pamphlet "Français, encore un effort si vous voulez être révolutionnaire" (in *La Philosophie dans le boudoir*). First the narrator asks why minor transgressions, such as masturbation, should be considered unnatural, since nature obviously encourages these acts. The reader assents. Not long after, we are being forced to accept sodomy, murder. Suddenly one feels the humiliating duplicity of one's continuing assent, if only in the act of continued reading. Nevertheless, we continue.

9. This intense imaginary identification with Christ's suffering is a Catholic tradition very much still in vogue today; one thinks of Mel Gibson's recent film *The Passion of the Christ,* which dwells almost exclusively on the extreme violence with which Jesus was tormented. One can imagine St. Ignatius of Loyola's reaction to the film.

10. Bataille's *ipse* is a nonself, the movement through which the traditional philosopher's *ipse*—a totalizing and solipsistic self—encounters and is lost in a "night" of "continuity." This movement doubles that of the "completion" of knowledge as non-knowledge.

11. Bataille at one point certainly thinks so. In the 1947 essay "From Existentialism to the Primacy of Economics," he writes: "I [Bataille] say personally: 'I know nothing, absolutely nothing. I cannot know *that which is.* Being unable to relate *that which is* to the known, I remain on the wrong path in the unknown.' *Inner Experience* expresses entirely this situation, which is that of the *there is [il y a]* of Lévinas, and to which the sentence in question in Blanchot gives an accomplished expression" (1999, 168; italics Bataille's; the sentence Bataille refers to is taken from Blanchot's *Thomas l'obscur*).

12. On Kojève's influence in prewar Paris, see Descombes, 21–70.

13. Kojève's reading of Hegel is an anthropology because its central dialectic is that of the master and slave: the end is thus not Hegelian Spirit, but Man—a slave, alive or dead, as the case may be. Mikkel Dufrenne notes (1948, 397) that Kojève's stress on finitude and mortality establishes his Hegelianism as a revisionary Heideggerianism. One could, of course, say the same for Bataille, who stresses human mortality and finitude in works such as the prewar "Obelisk" (see chapter 4 in this book). Mortality for Bataille and for Kojève, however, are very different things.

14. "Man" is used here in conformity with Kojève's usage; Kojève does not write of or mention women in his work. Man in a Kojèvian, posthumanist sense is the highest manifestation of the human, prior to the death of Man in Man's own accomplishment. Throughout this book I will use "Man" in the same humanist sense, all the while supposing that Man is not so easily killed off as Kojève would have it—and that his death involves more than the mere accomplishment of an albeit supreme task.

15. Kojève did not actually write his *Introduction à la lecture de Hegel* (1980; hereafter cited as *ILH*) specifically for publication; it was published in the late 1940s, compiled from lecture notes taken in the seminar some ten years before. Raymond Queneau compiled the notes, and Kojève checked them and added the footnotes. By the late 1940s he was already involved in administrative work for the

French government (and later the European Community), and, given the fact that he was implementing the end of history in the real world—the real, final work of philosophy—he had little time for teaching a dead discipline (the "end of philosophy" meant precisely that for Kojève).

16. Perhaps it is worth noting that Kojève capitalizes this term (like so many other words in his lexicon); Bataille, just as significantly, doesn't.

17. This latter point was noted by Derrida some time ago, in his article on Bataille: that the Bataillean movement is one of the incessant transfer of sense to non-sense, and the miming, thus parody, of philosophical concepts, which leads not to another truth, another concept, but to the evacuation of the concept. A circular agitation of the sign, in other words, where sense always reasserts itself in its periodic awareness of the primacy of its own non-sense. "Bataille, thus, can only utilize the *empty* form of the *Aufhebung,* in an analogical fashion, in order to designate, *as was never done before,* the transgressive relationship which links the world of meaning to the world of nonmeaning. This displacement is paradigmatic: within a form of writing, an intraphilosophical concept, the speculative concept par excellence, is forced to designate a movement which properly constitutes the excess of every possible philosopheme" (Derrida 1975, 275; italics Derrida's).

18. This could be characterized as a "good duality," the "left-handed" affirmation of sense and its loss, comparable to the "good duality" of economic production and senseless expenditure I discussed in chapter 2.

19. Sartre in fact continued his interest in (bad-faith) evil after the war with a book-length biography of Jean Genet *(Saint Genet, Comedian and Martyr,* 1963). Perhaps Sartre saw Genet as more of a threat, since he represented the next generation (Sartre's own, in fact)—whereas Bataille, for Sartre, was more emblematic of the prewar, surrealism-tinged generation.

20. For an interesting sociological explanation for Sartre's postwar success and Bataille's (continued) relative obscurity, see Boschetti 1988, 159–66.

21. See Williamson 2001, 281. For more on this question, see chapter 6.

22. Bataille comments in "From Existentialism to the Primacy of Economy": "Desire in Hegel is resolved thus in a knowledge that is absolute, that is a suppression of the subject who knows. One exists no longer in these conditions; history, first of all, is supposed to be completed, and so must be even the life of the individual subject. Never, if one thinks about it, has anyone conceived of anything more dead; the multiple life was the immense game and the immense error that the achievement of this death necessitated. Toward the end of his life, Hegel no longer posed the problem to himself: he repeated his lectures and played cards" (1999, 158).

4. Bataille's City

1. Sade's utopianism was, as he conceived it, a radical republicanism, hence generalized prostitution, quasi-legalized murder, lax book censorship (obviously), etc. Bataille too conceived a kind of utopian potlatch, first in the 1930s, in the con-

text of proletarian revolution (see, for example, "Popular Front in the Street" [*OC,* 1: 402–12; *VE,* 161–68]), then in terms of a secret society conspiracy (the Acéphale group [*OC,* 1: 442–46; *VE,* 178–81]), and finally in the context of a vast industrialized economy that engages in massive amounts of gift-giving (announced by the Marshall Plan, in *The Accursed Share*).

2. See, for example, Durkheim's "Individualism and the Intellectuals" (1973a) as well as *Moral Education* (1992).

3. Bataille parodied this secular humanistic ritual in the Acéphale period (the late 1930s): his group planned to put a skull softened in brine at the base of the Obelisk on the anniversary of Louis XVI's decapitation, proclaiming to the press that the king's skull had appeared. On Durkheim and the institution-representation of the community through totemic acts, see "Durkheim and the Totem Act," chapter 1 of my *Agonies of the Intellectual* (1992).

4. See preceding note.

5. In Sadean terminology, which Bataille no doubt would have approved, the "hot" (non)laws of the early, "criminal" Republic, exemplified by the execution of the king, were replaced by the "cold," tryrannical laws of the modern (Third) Republic. Needless to say, those "cold" laws were and are rigorously antisacrificial.

6. The definitive analysis in Bataille of the military leader and of fascism as embodiments of the right-hand sacred (heterogeneous or base elements reconfigured to serve as guarantors of political conservation, reactionary force, elevation, holiness, etc.) is to be found in "The Psychological Structure of Facism," especially *OC,* 1: 362–71; *VE,* 153–59. This essay is of great importance because it traces out the movement by which a radically unconditioned, literally meaningless element—for Bataille, the dead God—is reappropriated by society to serve as the capstone of all social meaning and the purpose, or end of all activity. Sovereignty indeed "is" an end, leading nowhere beyond itself; this is its appeal to social forces of construction, which could not act without an end. The problem is that in acting and laboring for the end, they are necessarily betraying it. The paradox in Bataille is that any constructive activity, any meaning (including Bataille's own writing) necessarily partakes of betrayal, of this fascistic moment. The only way out, if it is one, is in the movement of sense to non-sense, Bataille's procedure in the *Summa Atheologica.* Nevertheless, because of this inevitable reappropriation of sovereignty—a reappropriation associated in "The Psychological Structure" with fascism, there is in Bataille a certain, inevitable, guilty conscience and an irredeemable sense of complicity with fascism. See, for example, "Politics, Mutilation, Writing" (on Bataille's *Blue of Noon*) in my book *Politics, Writing, Mutilation* (1985).

7. In "The Pineal Eye," Bataille writes of the evolution/devolution of apes: "Because of the erect posture, the anal region ceased to form a protuberance, and it lost the 'privileged power of points': the erection could only be maintained on condition that a barrier of contracted muscles be regularly substituted for this 'power of points.' Thus the obscure vital thrusts were suddenly thrown back in the direction of the face and the cervical region" (*OC,* 2: 34; *VE,* 89). As in Durkheim, for

Bataille the sacred is constituted by a charge comparable to that of electricity: for Bataille, however, localized in or on the body, the charge moves around, disfiguring the human form in and as "evolution."

8. See a number of the essays in the Bataille collection *Visions of Excess,* such as "Base Materialism and Gnosticism," "The Big Toe," "Sacrificial Mutilation and the Severed Ear of Vincent van Gogh," etc.

9. It should be stressed that the fully sovereign figure in Bataille is, strictly speaking, impossible, since it is radically unconditioned, dependent on no one, subordinate to nothing. Writing of Sade's sovereign criminals/heroes, Bataille writes: "But de Sade's sovereign man has no actual sovereignty; he is a fictitious personage whose power is limited by no obligations" (*OC,* 10: 173–74; *E,* 174). None of Bataille's sovereigns, in fact, have "actual" sovereignty either, since any real-world sovereignty is necessarily limited. We can say instead that their sovereignty is "mythical," having the virulent force of base matter or (as in the case of Mme Edwarda) aroused organs.

10. See Guerlac 1990 for an analysis of "recognition by a woman" in Bataille, in the context of an all-encompassing fictionality.

11. "Prostituted" since prostitution entails nonreproductive sexuality, the squandering of sexual force and money, traditionally consecrated by religion. Prostitution in this sense is the "truth" of sexuality, since sexuality is only secondarily tied to reproduction; its primary movement is one of the expenditure of energy (and accumulated energy: wealth) and the movement of the being toward death through the destruction of unusable excess. Only as an unplanned aftereffect is it recuperated as reproduction: in the act of sexual congress, aroused beings are hardly concerned with the consequences of their acts. Once again, the right hand does not know—or is unconcerned with—what the left hand is doing. In *Erotism,* Bataille writes: "The prostitutes in contact with sacred things, in surroundings themselves sacred, had a sacredness comparable with that of priests. . . . We must not forget, however, that outside Christianity the religious and sacred nature of eroticism is shown in the full light of day, sacredness dominating shame" (*OC,* 10: 133–34; *E,* 133–34).

12. The Porte St. Denis was constructed by Louis XIV in 1673 to commemorate his victories on the Rhine. It once stood as one of the gates of the city, and the king passed under it when he entered the city, in royal procession, from the outlying town of St. Denis. Not only does it recall the acephalous St. Denis, who would have walked down this very street carrying his head in his hands, and the Roman god Dianus (Bataille's pseudonym as author of "The Torment," roughly contemporaneous with *Mme Edwarda*); it is decorated with martial emblems that represent suits of armor—without heads. The arch has recently been cleaned, and it looks almost new. Standing under the arch one can survey even today the busy street life of one of Paris's "hottest" quarters (although a recent crackdown on the owners of apartments rented out by the hour to prostitutes has put quite a few of them out of business). The "maisons closes" of Edwarda's day (hers was called "Les

Glaces"—"The Mirrors") are, however, long gone, abolished by the Marthe Richard law of 1946.

13. On Bataille's rewriting of Breton's "God, whom one does not describe, is a pig," see Hollier 1989, 29.

14. On the copula/copulus in Bataille, see ibid., 127.

15. See Hubbert 1949 for a fascinating discussion of the future decline of fossil fuel supplies. This essay (dating from 1949, the year of the publication of *The Accursed Share*) foresees not only the decline of easy energy availability, but argues for the urgent need to consider fuel alternatives. The essay is less concerned with the exact parameters of the depletion curve, mathematical models, etc., than it is with more profound questions having to do with the future of civilization. The author ascribes the "rise" of industrial civilization (and the concomitant global population increase) to the easy availability of a rich fossil fuel mix and the inevitable decline of that civilization to the eventual depletion of significant reserves of those fuels (108). What is striking is that things have not progressed much since Hubbert's time: nowadays other authors, such as Kunstler and Heinberg, are still calling for a massive effort to transition to new modes of energy consumption before it is too late—with the same lack of "traction" that Hubbert enjoyed in 1949.

5. Orgiastic Recycling

1. As Kojève puts it, "Now, several voyages of comparison made (between 1948 and 1958) to the United States and the USSR gave me the impression that if the Americans give the appearance of rich Sino-Soviets, it is because the Russians and the Chinese are only Americans who are still poor but are rapidly proceeding to get richer" (*ILH,* 436–37). Bataille as well assumed that the Americans and Soviets will converge at some point if they can keep from annihilating each other. Each will eventually evolve a kind of consumerist state socialism with, one assumes, a certain amount of individual freedom. If they are to do this, they will both have to engage in massive amounts of gift-giving—and in this the Soviets will have to follow the Americans' lead (in the Marshall Plan). This in any case is Bataille's argument in the last chapter of *The Accursed Share.*

2. For a sobering assessment of the depletion of the earth's resources—the mechanized "tendency to expend" and its consequences—see Brown 2001. On the consumption (and waste) of very specific (and surprisingly large) quantities of materials in the production of the simplest everyday objects and foods in consumer society, see Ryan 1997.

3. "Since economic activity requires work, and since energy is the capacity to perform work, energy is therefore necessary for economic growth" (Heinberg 2004, 37).

4. See Beisner 1997 on sustainability: "Given the dynamic character of human activity, it is difficult to see just what is meant by sustainability. Presumably... a procedure is sustainable if it can be continued indefinitely into the future. But

exceedingly few human procedures—technologies—continue unchanged even for a few generations, let alone into the indefinite future" (29–30).

5. This is the general argument of Heinberg, *The Party's Over* (2003). We should note that what is limited is not so much the energy available on the surface of the earth but rather the amount of energy supplies that lend themselves readily to "refinement"—that is, to their treatment by humans in such a way that they give up a large amount of *usable* energy—hence, again, the concept of EROEI (energy return on energy investment). This is the point at which an idealist line of interpretation is often asserted—that is, that human ingenuity, human spirit, will figure out new ways of producing harnessable energy from unpromising sources, because it always has (see Huber and Mills 2005). That, in effect, is the miracle of technology. This seemingly hard-core technophile argument is in fact a product of the purest form of humanist idealism: the human mind will always triumph because it always has (i.e., it has always managed to find new sources of energy). We could rewrite Heinberg's response to this sort of argument, in a Bataillean mode, in this way: matter is base; it is recalcitrant; it cannot always be made to wear a "frock coat." At some point the hubris of Man—the belief that the energy of the universe serves no other purpose than to be appropriable and "serve Man," that grotesque and phantasmic signifier—runs up against a profound barrier: there is an enormous amount of energy *that is not servile.* It spirals out of celestial bodies, it blows away in the wind, it courses uselessly through our bodies. Its expenditure is our finitude (our mortality), and its finitude—in the sense that heterogeneous energy opens the possibility but also defines the limits of the homogeneous energy that *can* serve—is our expenditure (our waste of effort, of time, of our own self-satisfaction).

6. The complete mantra was "Use it up, wear it out; make do, do without!" The postwar era in America was nothing more than the aggressively willed forgetting of this slogan, which, during the war, had been drilled into millions of heads at great government expense. As late as 2002 my mother still repeated it, with great satisfaction, as she stockpiled pencil stubs and scraps of envelopes in various cabinets.

7. Cited in Tierney 2004.

8. Thus the by-now (in)famous remark: "Agriculture is now a motorized food producing industry, in its essence the same thing as the fabrication of cadavers in the gas chambers and the concentration camps, the same thing as the blockade and reduction of countries to famine, the same thing as the fabrication of hydrogen bombs" (cited in Lacoue-Labarthe 1987, 58; my translation).

9. This is the subject of Michel Foucault's great book *Discipline and Punish* (1975): the process by which, in the nineteenth century, "Man" became an object of study, quantifiable, knowable, manipulable. Foucault's debt to Heidegger is evident.

10. As Bataille puts it in an early *(Documents)* article, "The Modern Mind and the Play of Transpositions," "What is truly loved is loved above all in shame and I challenge any painting fan *(amateur de peinture)* to love a canvas as much as a fetishist loves a shoe" (*OC,* 1: 273).

11. See, for example, James Howard Kunstler's remarks on Joel Kotkin's celebration of suburbia in his blog "The Clusterfuck Manifesto," 15 February 2005.

12. On the Derrida-Marion discussion, see especially Horner 1995.

13. Thus the Western model of prosperity spreads to formerly third world countries like China or India—and the result is that the depletion point, the peak of the curve, is only that much closer.

14. See Bataille's essay on the Kinsey report ("Kinsey, the Underworld, and Work"), for example (originally published in *Critique,* and reprinted in *Erotism* as chapter 1 of part 2, *OC,* 10: 149–63; *E,* 149–63). Bataille's subtitle could serve as the motto of my discussion of Heidegger and of the intimate world as opposed to the usable object: "Eroticism is an experience that cannot be assessed from outside us in the way an object can."

15. *Blue of Noon* and *L'Abbé C.* are both novels written by Bataille: the former in 1936 (though first published in 1957), the latter in 1949. They are reprinted in volume III of Bataille's *œuvres Complètes.* See also Bataille 1978 and Bataille 1983.

16. This is (part of) the definition of the word given in the 1974 edition of the *American Heritage Dictionary.*

17. For a trenchant critique of the contemporary French government's ineffective and socially destructive response to globalization, see Smith 2004.

18. The film presents a lawyer standing in a field who cites a French law permitting gleaning of already harvested fields—a law that can be directly traced back to the sixteenth century.

19. Much of Varda's film depicts people rummaging through garbage bins for food—as a contemporary, urban form of gleaning.

6. The Atheological Text

1. See the thought-provoking article by Perelman 1981, which, in its opening pages at least, suggests that a solar economy inevitably leads, due to the scarcity or feebleness of energy inputs, to a hierarchical, feudal economy with a concomitant, oppressive religion (social and spiritual hierarchy, right-hand, official religious practice, etc.). After this stimulating thesis, Perelman gets lost in considerations of social "schizophrenia" as a response to the energy crisis of the time. In his opening remarks Perelman might be on to something, but only in relation to agricultural societies. After all, hunter-gatherer societies are also solar-power-based, and they tend to be much more egalitarian (though, as LeBlanc would note, also quite violent). Some critics (e.g., Richard Manning) see a real fall in the movement from hunting-gathering to agriculture: the latter not only breeds feudalism, but organized and oppressive religion, social classes, ecocide, the oppression of women, weakness and malnutrition among the majority of the population, etc.

2. See for example, Blanchot: "A word gives me its meaning, but first it suppresses it. For me to be able to say, 'this woman,' I must somehow take her flesh and blood away from her, cause her to be absent, annihilate her. The word gives me the being, but it gives it to me deprived of being" (1995, 322).

3. One of the contributors to the *Daedalus* issue on ecoreligion is Sally McFague, a noted ecofeminist. For a trenchant critique of McFague's humanism, see Sideris 2003, chapter 2. The big question in ecohumanism remains: is a "community" of life forms, many of which are in violent competition with each other or which spend their lives confronting the fact of being hunted (zebra, for example), an appropriate model for a humanistic community?

4. Responding to the legal scholar Own Fiss, Fish writes: "More generally, whereas Fiss thinks that readers and texts are in need of constraints, I would say that they are structures of constraint, at once components of and agents in the larger structure of a field of practices, practices that are the content of whatever 'rules' one might identify as belonging to an enterprise" (1989, 133). For Fish, a text is an "already-interpreted object" (130).

5. But as a totem-act the flag is *empty*.

6. Thus Greg Bahnsen argues that we are "obligated" in all aspects of our lives by the laws of the Bible and even that, barring specific revision by New Testament law, we are fully bound by Old Testament law (Bahnsen 1985, 3). In a similar argument, theologian Rousas John Rushdoony argues that the basic law consists of the Ten Commandments, with all other Biblical passages consisting of application and interpretation (1973, 4–14). In both Bahnsen's and Rushdoony's views, Biblical law is unchanging, absolute, and the role of modern jurists consists of nothing more than literal application with an absolute minimum of interpretation. Even the New Testament, according to these authors, is concerned with only the most minor amending and interpreting.

7. It should be noted, however, that for Qutb the ultimate freedom is not that of freedom of conscience—a minor freedom—but freedom above all to practice Islam (Qutb 1978, 137). This logic is also seen among Christians: the traditionalist Catholic leader Monsignor Marcel Lefebvre writes, "*Subjective right* is the faculty of requirement, to the extent that it is rooted in the subject, regardless of its application; for example, the right to worship God, regardless of the content of that worship. *Objective right* is on the contrary the concrete object of the right: this worship, this education" (2000, 36).

8. Note that the supposedly erroneous predictions of depletion that Beisner (1997) cites, from as early as 1908, foretell a falloff in *U.S.* oil production (e.g., "1949: End of U.S. oil supply in sight"—is this Hubbert?). The irony here is that the sources quoted *were right:* U.S. oil production did peak in 1970. Thus the predictions were not as absurd as Beisner makes them out to be, and this in itself suggests that other predictions—of worldwide production peak—might also be accurate.

9. Reader-response theory, deconstruction, phenomenological criticism, even New Criticism all attempt to understand reading as a process that involves active, questioning agents who evolve an understanding of the text based on finite understanding. Reading then, is never absolute but is always conditioned by temporal, historical, community, and rhetorical limits. The literalist religious tradition alone (which includes, in this context, Kojève) dares to presuppose a monstrous text fixed,

once and for all, and necessarily understood in a single way through a single (infinitely repeated) act of reading.

10. A third tradition of reading, going back at least to St. Augustine, is that of allegorical reading. This is a tradition that relies on biblical interpretation and commentary—unlike the literalist tradition—and at the same time finds in the Bible one bedrock meaning beyond interpretation. Perhaps the modern avatar tradition is best represented today by current Vatican doctrine (see Williamson 2001), which holds that correct biblical reading requires both textual analysis (linguistic, rhetorical, literary reading) and a recognition that there is nevertheless one, unchangeable (and divine) religious doctrine conveyed through all the textual vagaries. While the Catholic Church certainly retains its authority today, one could argue that the two methods of biblical reading I have outlined in this chapter—pick and choose and literal—have recently become much more seductive, certainly in the context of ecological debate.

11. See Bataille: "But de Sade's sovereign man has no actual sovereignty; he is a fictitious personage limited by no obligations" (*OC,* 10: 173–74; *E,* 174). Of course, sovereignty for Bataille is not so much fictional as it is mythical; on the opposition (if it can be called that) between myth and fiction, see chapter 1 in this volume.

12. On recognition in and through sacrificial communication, Christopher M. Gemerchak has this to say: "Bataille affirms that beings communicate with their beyond through wounds alone, that they communicate with one another through the ruptures inflicted upon individual integrity. This is the path to (divine) intimacy, the annulment of the type of transcendence conceived as the separation between merely opposed—and thus reconcilable—individual/things, and the expansion of transcendence into a mutual absence where God is experienced" (2003, 165).

13. Luke sees a "soft anthropomorphism" lurking behind the Earth First! movement—certain humans, certain animals are valued over others. What's more, "While acknowledging that killing and suffering are natural or necessary, they [Earth First!ers] advance no criteria for deciding between alternatives when such 'natural acts' become necessary" (1997, 27). Were Luke a reader of Sade—perhaps the first rigorous (too rigorous) ecologist—he could argue that *any* hierarchy is an absurdity in the eyes of nature. That of course only leads to an ethics of murder, which one doubts the Earth First!ers would be happy to embrace. The problem, then, lies in evolving an ethics that inevitably privileges the human while at the same time recognizing that human populations are subject to the same constraints as all other animal populations: bloom, die-off, etc. Jared Diamond's book *Collapse* (2005) is another attempt at reconciling a view of humanity as just another (destructive, exotic) species in a given ecology with an ethics of the preservation of both human populations (and, one assumes, any given human individual) and larger ecologies (with all their indigenous species). Diamond's diagnoses of the failures of various societies (the Greenland Norse, the Easter Islanders, etc.) due to a kind of ecological hubris are compelling; his optimistic prognoses for the future much less so. It seems that human populations, like any other animal populations, tend to consume

as much energy as is available (bloom); when this sudden energy availability comes to an end (fragile soil, climate change, oil depletion), so does the population (die-off). Putting a happy spin on the possibilities for the current world population overhang would seem to have more to do with fantasy than with science (fiction). The real issue might be less the possibility of a pristine nature, either before or after a (soon to come) massive human mortality event than a consideration of the future of human-animal interaction *after* that die-off.

14. Perhaps David Abram suggests something along similar lines, although Abram does not stress the sacrificial aspect of the meeting of the human and the animal. It is, instead, primarily one of (non)recognition. Abram writes of meeting a "rare and beautiful" bison that exists only on Java: "Our eyes locked. When it [the bison] snorted, I snorted back; when it shifted its shoulders, I shifted my stance; when I tossed my head, it tossed *its* head in reply. I found myself caught in a non-verbal conversation with this Other, a gestural duet with which my conscious awareness had very little to do. It was as if my body in its actions was suddenly being motivated by a wisdom older than my thinking mind, as though it was held and moved by a logos, deeper than words, spoken by the Other's body, the trees, and the stony ground on which we stood" (1997, 21). Out of this duet Abram (following Merleau-Ponty) elaborates a theory of the body, of spirituality, and of ecology. Bataille, however, might hesitate at the notion of a "deeper" logos and of a sort of pantheism linking the body to the physical environment. His "communication" is one of a violent rupture rather than of a larger integration of an ultimately coherent body into a well-defined physical and spiritual universe. (Abram's face-off with the bison is actually quite reminiscent of a Sartrian freedom, which has its origin in the reciprocal recognition of human freedoms, often via the gaze. Abram's variation is that the other for him is "natural," an animal, rather than a socialized or socializable human.)

7. An Unknowable Future?

1. Just as God can be seen to be the ultimate constraint and meaning of gold for Goux, so too the ultimate constraint and meaning of the car is Man, the empty satisfied self.

2. On the desymbolization of urban space, see Augé 1992, especially "Le Lieu anthropologique" (57–95).

3. Both the chapter on de Certeau in Highmore 2002 and Brian Morris's article on "Walking in the City" (2004) provide excellent analyses of de Certeau's position and of the fixation on a rather simplistic "resistance" among de Certeau's commentators.

4. See Le Corbusier 1941. Le Corbusier and Bataille were both, in fact, energeticists. Le Corbusier approvingly cites Dr. Pierre Winter: "Human being is nothing but a transformer of solar energy; life is only the circulation of this energy; light is one of our fundamental nutrients *[aliments]*" (Le Corbusier 1941, 15; my translation). (We absorb it through our skin, we also draw each day from

the "reserves of light" constituted by most animal and plant-derived nutrients.) The difference between Le Corbusier and Bataille is, however, as great as their similarity. While Le Corbusier argued that urban life must be rigorously engineered, through the use of the automobile and the careful separation of social space in order to maximize the healthful absorption of solar light by each citizen, Bataille argued that solar energy is above all *expended* through disorganized, erotic, and anguished physical and intellectual movement that overturns the spurious ideals of God and Man.

5. As Huber and Mills put it, "As a general rule, more efficient devices are more efficient because they run *faster*" (2005, 114). Thus not driving fast in a car—riding somewhat more slowly on a bike—is wastage, as is the waste of resources that went into the construction and use of the car. The difference is that in the case of the bike the wastage is of the body—body effort in body time, destroying the idealized, totally efficient speed—which is also a form of stasis—of the car. The wastage of the car, in other words, is subordinate to an ideal of quantified efficiency (time is money), whereas the wastage of the bike is pointless, leads to nothing, entails an energy that cannot be used to maximum efficiency, maximum speed. The car is subordinate to an ideal meaning (that of money) that might itself seem unconditioned, but which is at the service of practical activity (it is thus, in Bataille's terminology in "The Psychological Structure of Fascism," an instance of "imperative heterogeneity").

6. What is more visually striking than a marginalized pedestrian or cyclist obstinately making headway alongside roaring traffic, in the rain?

7. In Goux's terminology (in "Numismatics" [1990b]), the car would be a general equivalent—an empty signifier serving as the possibility of the instantiation of meaning and value of all other things. It is excessive, and yet serves to anchor all other meanings, making the coherent organization of meaning possible. Thus the car serves to order and give meaning to all social life, career aspirations, pleasure, etc. Its value is as obvious, as transparent, as money. The self, on the other hand—Man as consciousness and subject—is what Goux calls a "counter-investment," "the surplus that subjugates" (1990b, 54, 61). This is a "surplus that exploits and enslaves." Man would share this character with (according to Goux) God, Logos, and the State. And Bataille might add, Law. Another classic example of a counter-investment, analyzed by Bataille, is the fascist dictator (in "The Psychological Structure of Fascism," 1970). At the top of the pyramid, these "counter-investments" return, in their emptiness, to open themselves, to reveal themselves in their absence: they are the holes in and through which we fall, through which we confront or "experience" our not-knowing, our finitude.

8. See the discussion of "good" and "bad" duality in chapter 2 in this volume.

9. On the distinction in de Certeau between "strategy" and "tactics," see above all Highmore 2002, 156–61.

10. When oil hits $150 a barrel (a modest estimate for, say, 2015, at least according to some "Peak Oil" experts), car culture will very quickly become a distant, albeit fond, memory. The notion of globalization, celebrated or excoriated by so many

postmodern critics, will seem equally arcane, since the quick and cheap transport of people and things and even information will come to an end. For further considerations on the end of the fossil fuel era, see Heinberg (2003), as well as the articles and polemics on the rather sobering Web site http://www.dieoff.org.

11. See Strenski 2002 on Durkheim as a thinker of sacrifice; and Richman 2002 on Bataille's connection to the tradition of Durkheim and on both Durkheim and Bataille as latter-day avatars of the French intellectual tradition that conceives society to be grounded in sacrificial expenditure.

12. For a systematic and thorough indictment of the official American automobile culture, in all its social and ecological ramifications, see Alvord 2000.

13. Standards more and more associated with the mass propagation and maintenance of obesity. The universalization of car culture has resulted in an epidemic of obesity: the energy one would otherwise spend on transporting oneself (walking, cycling) comes to be accumulated on the body in the form of fat. See ibid., 89–90. One of the forms of wealth (hoarded energy) expended in future acts of post-car-culture potlatch will therefore be the accumulated wattles of fat that decorate the bodies of so many modern individuals.

14. A fancy way of saying that with the eclipse of fossil fuel and the car, the city itself is transformed from a "machine for living" into an intimate world. The city becomes a panoply, a feast, a festival of junk that is there not just to be purchased and "used," but to be ogled, sniffed, desired, rejected, loved, dreamed. The physical act of walking, dancing, or cycling, if pushed to its (il)logical end, results in a city open to "tactics" (in de Certeau's sense) of "communication" (in Bataille's).

15. On the sociotechnical frame—a regime entailing both a given technology and the social implementation from which it is inseparable—see Rosen 2002, 174–78. The sociotechnical frame on which Rosen focuses is that of the bicycle. Rosen makes this trenchant observation: "A sustainability-centered sociotechnical frame of the bicycle will have to wrest the values of freedom and autonomy back from the sociotechnology of the automobile and re-integrate them within a wider conception of sustainable mobility. The sheer difficulty of this task can be demonstrated by trying to imagine a car-dependent relative, friend or colleague attaching the same sense of freedom he or she identifies with their automobile to public transportation, car-sharing, or multi-modal trips. One cause for optimism is that for cyclists these values are intrinsic to their modal choice" (175). Rosen is an unusual author in that he recognizes the importance of emotional, even ecstatic, experience ("freedom," "autonomy") in the choice or development of a sociotechnical frame. In other words, mere grim sustainability is not enough; in the future some "inner experience," to borrow Bataille's term, may very well come into play in the development of technical alternatives.

16. This is what Bataille himself did, in 1949, in *The Accursed Share*. The late 1940s and early 1950s were an era of seemingly insuperable shortage and gray constraint: food rationing, housing shortages, cities in ruins, accompanied by the perceived need for a communism that would restrain or eliminate the inequalities of society and its attendant waste and therefore do away with *shortage*. Sartre was the

hero of the day, the setter of the intellectual agenda, certainly not Bataille—and Sartre's existentialist theory was based on a theory of *lack* (lack remedied, of course, by human intervention, hard labor). Bataille's gesture—to proclaim that the central problem of the postwar era was not shortage but the glut of wealth with which humans were destroying themselves—was magnificently perverse.

17. In Varda's film *The Gleaners and I* (discussed in chapter 5 in this volume) not only are physical discards recycled but so too are the intelligence and wisdom of a young man whose skills are obviously not needed in a "strategic" economy of production-consumption. A scientist who holds a master's degree, Alain, instead of "working," spends his time teaching penniless immigrants, outside of the school/university system and for free.

18. The swerve, of course, is the characteristic movement of the atoms of Lucretius, the author of the first radically materialist text. The swerve is the movement out of which arises (or falls) the complexity, freedom, and incomprehensibility of the world. Without the swerve, there would only be ideal inexistence: a fall of atoms without contamination. It is only when they interfere with each other that things happen, come into being, degrade. See Lucretius 1995, 63–65; and Vitzthum 1995, 37–38.

19. The bike here, like the table on which Bataille drinks wine (*La Limite de l'utile*, *OC*, 7: 344–45) is a practical object defeated in its practicality through its misuse (the flagrant negation of the purposive activity that went into its construction), in an "intimate moment" *(instant intime)*. The object is used, or used up, precisely in a gesture that defeats utility. Such a gesture, not coincidentally, is parodic, like much of the "recycling" we see in Varda's *The Gleaners and I* or like Bataille's doubling of the Hegelian-Kojèvian dialectic (in non-knowledge). The origin of the thing is defeated not frontally but in a swerving doubling and repetition of use as nonuse.

20. A similar movement of orgiastic (and parodic) display and expenditure combined with the affirmation of an ecologically sound alternative mode of (urban) transport can be seen in the case of Italian Critical Mass riders, who parade nude through the streets of various cities, their bodies spectacularly and suggestively painted. See http://www.inventati.org/criticalmass/ciclonudista/bodypaint.htm.

Critical Mass as practiced throughout the world in fact seems not so much about "doing" or "accomplishing" something through an "organization," but rather a (dis)organized celebration of another way of living, moving, and thriving in public space. A celebration that may set off change (if it does) through the example of people living differently and riotously. See Carlsson 2002.

21. It would seem to constitute the empty space of the universal, playing its role while leaving the universal blank—as in Laclau, a space of contact, communication, and rupture, rather than plenitude. Hardt and Negri avoid, however, the term "empty universal."

22. Of course I use the terms "heterogeneous" and "homogeneous" in the sense given by Bataille in "The Psychological Structure of Fascism" (*OC*, 1: 339–71; *VE*, 137–60). "Homogeneous" energy is energy that has been refined, made

efficient, and that is fully the "power to do work" under the control and direction of Man. "Heterogeneous" energy precisely resists refining, makes itself felt in ways humans cannot control, vents itself in scandalous and destructive ways. It cannot be easily ordered. It is both too "diffuse" (energy that can never quite be made to serve) and at the same time too dense, too sacred. It is the "power of points" migrating throughout the aroused body, agitating its cursed matter. Its immediate expenditure in and through the *ipse,* the self doomed to communication, cannot for that reason be purposive. It is the movement of the act in the instant, not the planned work harnessed to another goal, a transcendent heterogeneity (God, the leader, etc.) made homogeneous through its servitude to, and justification of, human demands. We might say that energy in its largest sense is neither useful nor useless, neither conserved nor wasted (since that presupposes a universe subordinate to Man and his needs), neither work nor idleness, but instead that it opens the possibility of those terms without ever being simply classable under one or the other. It is formless in relation to them. It not only defies useful, quantifiable labor; it defies the system of oppositions that would measure it, comprehend it, and put it to work. As such it is sovereign and charges bodies and the city, moving them without allowing itself to be easily comprehended (that is, it allows itself to be comprehended only in myth, a knowledge that is the defeat of knowledge).

23. Put another way, "Nature" as "source" of energy, in the very finitude of the usable fossil fuel energy supply, is heterogeneous to human needs and even understanding. "Nature" is the death of the plenitude of modern self-understanding (the sheer presence of the motorized "I," caught in the eternity of its present). On the (non)relation between self and nature, especially in theories of the sublime, see Martyn 2003, chapter 5 (on Kant's *Critique of Judgment*), 135–69.

24. See ibid., 140.

25. See John C. Ryan's book *Stuff* (1997), which very concretely presents and analyzes the energy and resource inputs of everyday objects—a cup of coffee, a newspaper, a pair of shoes, etc. This is a very sobering work and a good indication of the power of straightforward analysis (accounting); what is most remarkable is the extent to which this information is hidden from us, or willfully ignored by us, in our daily lives.

26. Empires (in the common usage of the word) are associated with specific energy regimes: the empire of Genghis Khan was grass fed, solar, and was based on land (the horses of his warriors derived their energy from the lands they traversed); the British empire in its heyday was wind powered and based on the sea (British ships depended on sails to make possible the cohesion of the empire); the American empire is fossil fuel powered and based in the air (fighter jets, drone surveillance planes, passenger jets, etc.).

27. The dollar no longer represents a certain quantity of gold, but so long as its value in barrels of oil is relatively stable, it is still worth something. The price of oil serves as a limit to the tendency toward hyperinflation, the result of massive indebtedness and the consequent printing of ever more dollars. When the price of oil skyrockets, however, the value of the dollar will rapidly evaporate.

28. William Kunstler's blog ("The Clusterfuck Manifesto"), May 16, 2005, notes with some astonishment the fact that no one is concerned with imminent fossil fuel depletion at a major car show. Of course car cultists would not be—but this leads to the sobering realization that *anyone* who drives must also remain in a state of willed blissful ignorance concerning the fate of fossil fuel, since anyone who drives necessarily believes in the car as a solution and not as a problem.

29. Thus Huber and Mills (2005, *The Bottomless Well*): human spirit always finds solutions to energy crises. Note that the idea that there will never be an energy crisis is closely tied to idealism: the human mind (not simply human labor as in Marxism) will alone provide an infinite amount of energy via the miracle of technology.

30. Price, as well as a number of other articles at the http://www.dieoff.org site, stresses the incontrovertible link between population growth and the availability of energy resources (and the corresponding fall in population with a decline in energy resources).

31. It should be stressed that Bataille seemed to have a double take on the affirmation of a Kojèvian absolute knowledge. The first, which I discussed in chapter 3, affirms absolute knowing as a variant of non-knowledge *(non-savoir).* In other words, absolute knowledge is parodic, circular, empty to the extent that its knowledge is only knowledge of its own impossibility, its death. Elsewhere, however, Bataille affirms absolute knowledge on the social plane as the end of history. In this sense he affirms a Kojèvian historical moment where, presumably, all revolutions have taken place, all social knowledge is complete, and negativity truly is "out of a job." Along with this, presumably, would come the Hegelian-Kojèvian text that is beyond revision, beyond interpretation if not reading. In this instance, however, Bataille does not note how his own text, that of the *Summa Atheologica,* would fare, locked as a double in a constant struggle with the Kojèvian text. Nevertheless, Bataille does affirm in this context the march of modernity and its full accomplishment. Only at and after the Kojèvian end of history, in this version of things, can expenditure be recognized as a social force, as an "unemployed negativity." See Bataille's letter to Kojève in *Guilty* ("Letter to Blank, Instructor of a Course on Hegel," *OC,* 5: 369–71; *Guilty,* 1988b, 123–25). The letter dates from 6 December 1937—well before the positions of the *Summa Atheologica* or of the *Accursed Share.* Bataille to my knowledge never rigorously formulated how the ethics and political positions implied in *The Accursed Share* would alter the definitive knowledge embodied by and in a posthistorical state (if at all). If negativity is truly "unemployed," how could its implementation influence a posthistorical, posthuman regime? The danger of Bataille's position in the "Letter" is that his "unemployed negativity" would be indistinguishable from the negativity that Kojève sees manifesting itself at and after the end: the purely formal negativity best exemplified by the Japanese.

32. See, again, the "Letter to Blank" (*OC,* 5: 369–71; *Guilty,* 1988b, 23–25).

33. For a trenchant critique of the ideology of fossil fuel culture, with a call to rethink the logic of resource conservation, see Tubbs 2005.

34. On energy and order, see Huber and Mills 2005, 23.

35. It can, however, work the other way: when one has slaves, one sees no need for supplemental energy derived from fuels. It is possible that the ancient Romans did not develop the steam engine (even though it was invented in antiquity by Hero of Alexandria) because they saw no need for the power of which it was capable: they had, after all, gangs of slaves, and fuel (large amounts of firewood, coal) was relatively hard to obtain. Of course labor was cheap in the ancient world (workers were slaves); under capitalism, labor has to be paid for, and the incentive to derive energy from wood and, ultimately, inanimate sources is much greater. See on this the entry for Hero of Alexandria in Eco 1963. On the invention of the steam engine as a device to facilitate the mining of coal—an engine powered by the very fuel whose mining it was facilitating—see the excellent and highly thought-provoking book by Freese (2003).

36. On the relation between threat and knowledge, see van Wyck: "The challenge of the real of ecological threats is precisely to discover a mediator that will allow something new to be said, that will perhaps allow *a qualitatively new manner of thought and action* to inform a time (ours, for instance) in which the productive capacity of threats seem to outstrip any reasonable capacity for reflective (affective) response" (2005, 112; italics added).

37. Which is not to say that conservation, conventional recycling, of materials, etc., are not desirable and worthwhile goals. My point, however, is that they will follow directly from a rethinking of expenditure: what society will end up having to affirm, as I have tried to argue, is not simple waste—fossil fueled production and destruction—but the burn-off of energy in the movement of an intimate world (muscles in agony and ecstasy, for example). If society's "freedom" is reoriented toward such an expenditure, then the other freedom, that of simply consuming *(consommer),* can be left behind. If such a reorientation can be affirmed, then conventional recycling can be something more, or other, than a mere sign of "personal virtue." The most rigorous recycling, then, may very well be the aftereffect of a "spending without return."

38. Perhaps a renewal of the sacred of the sort I am discussing in the post–fossil fuel era has already commenced. This is in any case the argument of Maffesoli (2004). A greater general awareness of the tension between the sacred (or the tragic) and quantifiable expenditure is, however, to be hoped for, and I think it will be an inevitable aspect of life in what Kunstler (2005b) calls the "long emergency."

Bibliography

Abram, David. 1997. *The Spell of the Sensuous.* New York: Random House/Vintage.

Alvord, Katie. 2000. *Divorce Your Car! Ending the Love Affair with the Automobile.* Gabriola Island, BC: New Society.

Ambrosino, Georges. 1950. "La Machine Savante et la vie: Norbert Wiener, *Cybernetics.*" *Critique* 41: 70–82.

Augé, Marc. 1992. *Non-lieux: Introduction à une anthropologie de la modernité.* Paris: Seuil.

Bahnsen, Greg. 1985. *By This Standard: The Authority of God's Law Today.* Tyler, Texas: Institute for Christian Economics.

Bataille, Georges. 1955a. *Lascaux, or The Birth of Art.* Trans. Austryn Wainhouse. Lausanne: Skira.

———. 1955b. *Manet.* Trans. Austryn Wainhouse and James Emmons. Lausanne: Skira.

———. 1970. "Psychological Structure of Fascism." In *Œuvres complètes* (1970–88), 1: 339–71; and in *Visions of Excess* (1985), 137–60.

———. 1970–88. *Œuvres complètes (OC).* Vols. 1–12. Paris: Gallimard.

———. 1977. *Story of the Eye.* Trans. Joachim Neugroschel. New York: Urizen Books.

———. 1978. *Blue of Noon.* Trans. Harry Mathews. New York: Urizen Books.

———. 1983. *L'Abbé C.* Trans. Philip Facey. New York and London: Marion Boyars.

———. 1985. *Visions of Excess: Selected Writings, 1927–1939 (VE).* Ed. A. Stoekl, trans. A. Stoekl, C. Lovitt, D. M. Leslie Jr. Minneapolis: University of Minnesota Press.

———. 1987. *Erotism: Death and Sensuality.* Trans. Mary Dalwood. San Francisco: City Lights.

———. 1988a. *The Accursed Share: An Essay on General Economy (AS).* Vol. 1, *Consumption.* Trans. Robert Hurley. New York: Zone.

———. 1988b. *Guilty.* Trans. Bruce Boone. Venice, Calif.: Lapis Press.

———. 1988c. *Inner Experience.* Trans. and introduction Leslie Ann Boldt. Albany: State University of New York Press.

———. 1992. *Theory of Religion.* Trans. Robert Hurley. New York: Zone.

———. 1994. *On Nietzsche.* Trans. Bruce Boone. Introduction by Sylvère Lotringer. New York: Paragon House.

———. 1999. "From Existentialism to the Primacy of Economy." Trans. Jill Robbins. In *Altered Reading: Levinas and Literature,* Jill Robbins, 155–80. Chicago: University of Chicago Press.

———. 2001. "Friendship." Trans. Hager Weslati. *Parallax* 18: 3–15.

Beaudreau, Bernard C. 1999. *Energy and the Rise and Fall of Political Economy.* Westport, Conn.: Greenwood Press.

Beisner, E. Calvin. 1997. *Where Garden Meets Wilderness: Evangelical Entry into the Environmental Debate.* Grand Rapids, Mich.: W. B. Eerdmans.

Blanchot, Maurice. 1949. *La part du feu.* Paris: Gallimard.

———. 1963. "La raison de Sade." In *Lautréamont et Sade,* 17–49. Paris: Minuit.

———. 1971. *Faux pas.* Paris: Gallimard.

———. 1992. *The Infinite Conversation.* Trans. Susan Hanson. Minneapolis: University of Minnesota Press.

———. 1995. *The Work of Fire.* Trans. Charlotte Mandell. Stanford, Calif.: Stanford University Press.

Blood, Susan. 2002. "The Poetics of Expenditure." *MLN* 117: 836–57.

Bois, Yve-Alain. 1997. *Formless: A User's Guide.* New York: Zone.

Boschetti, Anna. 1988. *The Intellectual Enterprise: Sartre and Les Temps Modernes.* Evanston, Ill.: Northwestern University Press.

Boyden, Stephen. 1987. *Western Civilization in Biological Perspective.* Oxford, UK: Clarendon Press.

Brooks, David. 2004. *On Paradise Drive: How We Live Now (and Always Have) in the Future Tense.* New York: Simon and Schuster.

Brown, Lester. 2001. *Eco-Economy: Building an Economy for the Earth.* New York: Norton.

Bruno, Giordano. 1975. *The Ash Wednesday Supper.* Trans. and with introduction by Stanley L. Jaki. The Hague, Netherlands: Mouton.

———. 1998. *Cause, Principle, and Unity.* Ed. and trans. R. J. Blackwell. Cambridge: Cambridge University Press.

Carlsson, Chris, ed. 2002. *Critical Mass: Bicycling's Defiant Celebration.* Oakland, Calif.: AK Press.

Clarke, Bruce. 2001. *Energy Forms: Allegory and Science in the Era of Classical Thermodynamics.* Ann Arbor: University of Michigan Press.

Connor, Peter Tracey. 2000. *Georges Bataille and the Mysticism of Sin.* Baltimore, Md.: Johns Hopkins University Press.

Costanza, Robert. 2006. "The Real Economy." *Vermont Commons* 10: 6–7.

Cronon, William. 1991. *Nature's Metropolis: Chicago and the Great West.* New York: W. W. Norton.

de Certeau, Michel. 1980. "Marches dans la ville." In *L'Invention du quotidien,* vol. 1, *Arts de faire,* 171–98. Paris: Bourgois, collection 10/18.

Deffeyes, Kenneth S. 2001. *Hubbert's Peak: The Impending World Oil Shortage.* Princeton, N.J.: Princeton University Press.

Derrida, Jacques. 1978. "From Restricted to General Economy: A Hegelianism without Reserve." In *Writing and Difference,* trans. Allan Bass, 251–79. Chicago: University of Chicago Press.

———. 1995. *The Gift of Death.* Trans. David Wills. Chicago: University of Chicago Press.

Descombes, Vincent. 1980. *Modern French Philosophy.* Trans. L. Scott-Fox and J. M. Harding. Cambridge: Cambridge University Press.

Diamond, Jared. 2005. *Collapse: How Societies Chose to Fail or Succeed.* New York: Viking.

Dufrenne, Mikkel. 1948. "Actualité de Hegel." Review of Kojève, *Introduction à la lecture de Hegel,* and Hyppolite, *Genèse et structure de la Phénoménologie de l'esprit chez Hegel. Esprit* 16, no. 9: 396–408.

Durkheim, Émile. 1973a. "Individualism and the Intellectuals." Trans M. Traugott. In *Emile Durkheim: On Morality and Society.* Chicago: University of Chicago Press. First published in 1898.

———. 1973b. *Moral Education.* Trans. E. K. Wilson and H. Schnurer. New York: Free Press. First published in 1902–3.

———. 1995. *The Elementary Forms of the Religious Life.* Trans. Karen E. Fields. New York: Free Press. First published in 1912.

Eco, Umberto. 1963. *The Picture History of Inventions from Plough to Polaris.* London: Macmillan.

Fish, Stanley. 1989. *Doing What Comes Naturally: Change, Rhetoric, and the Practice of Theory in Literary and Legal Studies.* Durham, N.C.: Duke University Press.

Foucault, Michel. 1995. *Discipline and Punish: The Birth of the Prison.* Trans. Alan Sheridan. New York: Vintage.

Freese, Barbara. 2003. *Coal: A Human History.* New York: Penguin.

Fukuyama, Francis. 1992. *The End of History and the Last Man.* New York: Avon.

Gardner, Gary. 2002. *Invoking the Spirit: Religion and Spirituality in the Quest for a Sustainable World.* Washington, D.C.: Worldwatch Institute.

Gatti, Hilary. 1999. *Giordano Bruno and Renaissance Science.* Ithaca, N.Y.: Cornell University Press.

Gemerchak, Christopher M. 2003. *The Sunday of the Negative: Reading Bataille Reading Hegel.* Albany: State University of New York Press.

Gottfrieb, Robert S. 1985. *The Black Death.* New York: Free Press.

Goux, Jean-Joseph. 1990a. "General Economics and Postmodern Capitalism." *Yale French Studies* 78: 206–24.

———. 1990b. "Numismatics: An Essay in Theoretical Numismatics." In *Symbolic Economies: After Marx and Freud,* 9–63. Ithaca, N.Y.: Cornell University Press.

Guerlac, Suzanne. 1990. "'Recognition' by a Woman! A Reading of Bataille's *L'erotisme.*" *Yale French Studies* 78: 90–105.

Hardt, Michael, and Antonio Negri. 2000. *Empire.* Cambridge, Mass.: Harvard University Press.

Heidegger, Martin. 1977. *The Question concerning Technology, and Other Essays.* Trans. William Lovitt. New York: Harper Torchbooks.

———. 1996. *Being and Time.* Trans. Joan Stambaugh. Albany: State University of New York Press.

Heinberg, Richard. 2003. *The Party's Over: Oil, War, and the Fate of Industrial Societies.* Gabriola Island, BC: New Society.

———. 2004. *Powerdown: Options and Actions for a Post-Carbon World.* Gabriola Island, BC: New Society.

Highmore, Ben. 2002. *Everyday Life and Cultural Theory: An Introduction.* London and New York: Routledge.

Hochroth, Lysa. 1995. "The Scientific Imperative: Improductive Expenditure and Energeticism." *Configurations* 3, no. 1: 47–77.

Hollier, Denis. 1989. *Against Architecture: The Writings of Georges Bataille.* Trans. Betsy Wing. Cambridge, Mass.: MIT Press.

———. 1990. "The Dualist Materialism of Georges Bataille." *Yale French Studies* 78: 124–39.

Hollywood, Amy. 2002. *Sensible Ecstasy: Mysticism, Sexual Difference, and the Demands of History.* Chicago: University of Chicago Press.

Horner, Robyn. 1995. *Rethinking God as Gift: Marion, Derrida, and the Limits of Phenomenology.* New York: Fordham University Press.

Hubbert, M. King. 1949. "Energy from Fossil Fuels." *Science* 109, no. 2823: 103–9.

Huber, Peter W., and Mark P. Mills. 2005. *The Bottomless Well: The Twilight of Fuel, the Virtue of Waste, and Why We Will Never Run Out of Energy.* New York: Basic Books.

Illich, Ivan. 1973. *Tools for Conviviality.* New York: Harper and Row.

———. 2001. *Energy and Equity.* London: Marion Boyars.

Karsenti, Bruno. 1997. *L'homme total: Sociologie, anthropologie, et philosophie chez Marcel Mauss.* Paris: PUF.

Kay, Jane Holtz. 1998. *Asphalt Nation: How the Automobile Took Over America, and How We Can Take It Back.* Berkeley: University of California Press.

Kerouac, Jack. 1958. *On the Road.* New York: New American Library.

Klossowski, Pierre. 1967. *Sade mon prochain.* Paris: Seuil.

Kojève, Alexandre. 1980. *Introduction à la lecture de Hegel (ILH).* Ed. R. Queneau. Paris: Gallimard, Collection "Tel."

Kunstler, James Howard. 2005a. "The Clusterfuck Chronicle." Blog. http://www.kunstler.com.

———. 2005b. *The Long Emergency.* New York: Atlantic Monthly Press.

Lacoue-Labarthe, Philippe. 1987. *La Fiction du politique.* Paris: Christian Bourgois.

LeBlanc, Steven A. 2003. *Constant Battles: The Myth of the Peaceful, Noble Savage.* New York: St. Martin's Press.

Le Corbusier. 1941. *Destin de Paris.* Paris: Sorlot.

Lefebvre, Marcel. 2000. *Mes doutes sur la liberté religieuse.* Paris: Clovis.

Lingis, Alphonso. 1989. *Deathbound Subjectivity.* Bloomington: Indiana University Press.

———. 1994. *The Community of Those Who Have Nothing in Common.* Bloomington: Indiana University Press.

———. 2004. *Trust.* Minneapolis: University of Minnesota Press.

Lomasky, Loren E. 1995. "Autonomy and Automobility." http://www.cei.org/pdf/1437.pdf.

Lucretius. 1995. *On the Nature of Things.* Trans. and introduction by Anthony M. Esolen. Baltimore, Md.: Johns Hopkins University Press.

Luke, Timothy W. 1997. *Ecocritique: Contesting the Politics of Nature, Economy, and Culture.* Minneapolis: University of Minnesota Press.

Maffesoli, Michel. 2004. "The Return of the Tragic in Postmodern Societies." *New Literary History* 75, no. 1: 133–49.

Manning, Richard. 2004. *Against the Grain: How Agriculture Has Hijacked Civilization.* New York: North Point Press.

Martyn, David. 2003. *Sublime Failures: The Ethics of Kant and Sade.* Detroit: Wayne State University Press.

Mauss, Marcel. 1967. *The Gift: Forms and Functions of Exchange in Archaic Societies.* Trans. Ian Cunnison. New York: W. W. Norton.

McKibben, Bill. 2001. "Where Do We Go from Here?" *Daedalus* 130, no. 4: 301–6.

Morris, Brian. 2004. "What We Talk about When We Talk about 'Walking in the City.'" *Cultural Studies* 18, no. 5: 675–97.

Mulder, Kenneth, Robert Costanza, and Jon Erickson. 2006. "The Contribution of Built, Human, Social, and Natural Capital to Quality of Life in Intentional and Unintentional Communities." *Ecological Economics* 59, no. 1: 13–23.

Naess, Arne. 1995a. "Deep Ecology and Lifestyle." In *Deep Ecology for the Twenty-First Century,* ed. George Sessions, 259–64. Boston: Shambhala.

———. 1995b. "The Shallow and the Deep: Long-Range Ecology Movements, a Summary." In *Deep Ecology for the Twenty-First Century,* ed. George Sessions, 151–55. Boston: Shambhala.

Newton, Lisa H. 2003. *Ethics and Sustainability: Sustainable Development and the Moral Life.* Upper Saddle River, N.J.: Prentice-Hall.

Odum, Howard T., and Elizabeth C. Odum. 2001. *A Prosperous Way Down: Principles and Policies.* Boulder, Colo.: University Press of Colorado.

Onfray, Michel. 2005. *Traité d'athéologie: Physique de la métaphysique.* Paris: Grasset.

O'Toole, Randal. 2001. *The Vanishing Automobile: How Smart Growth Will Harm American Cities.* Portland, Ore.: Thoreau Institute.

Perelman, Lewis J. 1981. "Speculations on the Transition to Sustainable Energy." In *Energy Transitions: Long-Term Perspectives,* ed. L. J. Perelman, A. W. Giebelhaus, and M. D. Yokell, 185–216. Boulder, Colo.: Westview Press.

Price, David. 1995. "Energy and Human Evolution." *Population and Environment: A Journal of Interdisciplinary Studies* 16, no. 4: 301–19.

Queneau, Raymond. 1951. *Le dimanche de la vie.* Paris: Gallimard.

Qutb, Sayyid. 1978. *Milestones.* Beirut and Damascus: Holy Koran Publishing House.

Richman, Michele. 2002. *Sacred Revolutions: Durkheim and the Collège de Sociologie.* Minneapolis: University of Minnesota Press.

Rosen, Paul. 2002. *Framing Production: Technology, Culture, and Change in the British Bicycle Industry.* Cambridge, Mass.: MIT Press.

Rushdoony, Rousas John. 1973. *The Institutes of Biblical Law.* N.p.: Presbyterian and Reformed Publishing.

Ryan, John C. 1997. *Stuff: The Secret Lives of Everyday Things.* Seattle, Wash.: Northwest Environment Watch.

Sade, D. A. F. de. 1966. *Histoire de Juliette, ou Les prospérités du vice.* In *œuvres Complètes,* vols. 9–10. Paris: Au cercle du livre précieux.

———. 1968. *La philosophie dans le boudoir.* In *œuvres complètes,* vol. 25. Paris: Pauvert.

Sartre, Jean-Paul. 1975. "Un Nouveau mystique." In *Situations I, Critiques littéraires.* Paris: Gallimard, collection Idées.

———. 1963. *Saint Genet, Actor and Martyr.* New York: New American Library.

Schumacher, E. F. 1989. *Small Is Beautiful: Economics As If People Mattered.* New York: Perennial Library. First published in 1973.

Sichère, Bernard. 1982. "Lècriture souveraine de Georges Bataille." *Tel Quel* 93: 58–75.

Sideris, Lisa H. 2003. *Environmental Ethics, Ecological Theology, and Natural Selection.* New York: Columbia University Press.

Smith, Timothy B. 2004. *France in Crisis: Welfare, Inequality, and Globalization since 1980.* Cambridge: Cambridge University Press.

Stoekl, Allan. 1985. *Politics, Writing, Mutilation: The Cases of Bataille, Blanchot, Roussel, Leiris, and Ponge.* Minneapolis: University of Minnesota Press.

———. 1992. *Agonies of the Intellectual: Commitment, Subjectivity, and the Performative in the Twentieth-Century French Tradition.* Lincoln: University of Nebraska Press.

Strenski, Ivan. 2002. *Contesting Sacrifice: Religion, Nationalism, and Social Thought in France.* Chicago: University of Chicago Press.

Tierney, John. 2004. "The Autonomist Manifesto, or How I Learned to Stop Worrying and Love the Road." *New York Times Magazine,* 16 September 2004, 58–65.

Trainer, F. E. 1995. *Consumer Society: Alternatives for Sustainability.* Sydney, Australia: Zed.

Tubbs, Nigel. 2005. "Fossil Fuel Culture." *Parallax* 11, no. 4: 104–15.

Tucker, Mary Evelyn. 2003. *Worldly Wonder: Religions Enter Their Ecological Phase.* La Salle, Ill.: Open Court.

van Wyck, Peter C. 2005. *Signs of Danger: Waste, Trauma, and Nuclear Threat.* Minneapolis: University of Minnesota Press.

Varda, Agnès, dir. 2000. *The Gleaners and I.* Film. New York: Zeitgeist Video.

Virilio, Paul. 1993. *L'art du moteur.* Paris: Galilée.

Vitzthum, Richard C. 1995. *Materialism: An Affirmative History and Definition.* Amherst, N.Y.: Prometheus Books.

Warman, Caroline. 2002. *Sade: From Materialism to Pornography.* Oxford: Voltaire Foundation.

White, Lynne, Jr. 1968. "The Historical Roots of Our Ecologic Crisis." In *Machina ex Deo: Essays in the Dynamism of Western Culture,* 75–94. Cambridge, Mass.: MIT Press.

White, Michael. 2002. *The Pope and the Heretic.* New York: William Morrow.

Williamson, Peter S. 2001. *Catholic Principles for Interpreting Scripture: A Study of the Pontifical Biblical Commission's "The Interpretation of the Bible in the Church."* Rome: Editrice Pontificio Istituto Romano.

Winnubst, Shannon. 2007. "Bataille's Queer Pleasures: The Universe as Spider or Spit." In *Reading Bataille Now,* ed. S. Winnubst. Bloomington: Indiana University Press.

Yates, Frances A. 1964. *Giordano Bruno and the Hermetic Tradition.* Chicago: University of Chicago Press.

Zabel, Graham. 2000. "Population and Energy." At http://www.dieoff.org.

Zeidan, David. 2003. *The Resurgence of Religion: A Comparative Study of Selected Themes in Christian and Islamic Fundamentalist Discourses.* Leiden, Netherlands: Brill.

Index

Allan Stoekl is professor of French and comparative literature at Penn State University. His interests include twentieth-century French intellectual history, contemporary literary theory, theories of energy, and the history of energy production and use. He is the author of *Politics, Writing, Mutilation: The Cases of Bataille, Roussel, Leiris, and Ponge* and the translator of *Visions of Excess: Selected Writings of Georges Bataille, 1927–1939*, both published by the University of Minnesota Press.